the River

Patrice Newell was born in Adelaide in 1956 and now lives in Gundy, New South Wales. She is the author of *The Olive Grove* (Penguin, 2000).

the River

PATRICE NEWELL

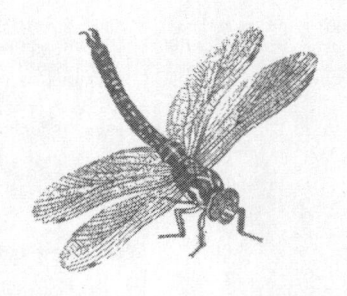

PENGUIN BOOKS

Penguin Books

Published by the Penguin Group (Australia)
250 Camberwell Road, Camberwell, Victoria 3124, Australia
Penguin Books Ltd
80 Strand, London WC2R 0RL, England
Penguin Putnam Inc.
375 Hudson Street, New York, New York 10014, USA
Penguin Books, a division of Pearson Canada
10 Alcorn Avenue, Toronto, Ontario, Canada M4V 3B2
Penguin Books (NZ) Ltd
Cnr Rosedale and Airborne Roads, Albany, Auckland, New Zealand
Penguin Books (South Africa) (Pty) Ltd
24 Sturdee Avenue, Rosebank, Johannesburg 2196, South Africa
Penguin Books India (P) Ltd
11, Community Centre, Panchsheel Park, New Delhi 110 017, India

First published by Penguin Books Australia, a division of Pearson Australia Group, 2003

1 3 5 7 9 10 8 6 4 2

Text copyright © Patrice Newell 2003
The moral right of the author has been asserted

All rights reserved. Without limiting the rights under copyright reserved above, no part of this publication may be reproduced, stored in or introduced into a retrieval system, or transmitted, in any form or by any means (electronic, mechanical, photocopying, recording or otherwise), without the prior written permission of both the copyright owner and the above publisher of this book.

Text and cover design by Nikki Townsend © Penguin Group (Australia)
Typeset in Berkeley Book by Post Pre-press Group, Brisbane, Queensland
Printed and bound in Australia by McPherson's Printing Group, Maryborough, Victoria

National Library of Australia
Cataloguing-in-Publication data:

Newell, Patrice, 1956– .
The river.

Includes index.
ISBN 0 14 300092 6.

1. Newell, Patrice, 1956– . 2. Pages River (N.S.W.) – Description and travel.
3. Pages River (N.S.W.) – History. 4. Water-supply – New South Wales – Pages River Region.
5. Human ecology – New South Wales – Hunter Valley Region.
I. Title.

919.442

This project has been assisted by the Commonwealth Government through the Australia Council, its arts funding and advisory body.

www.penguin.com.au

CONTENTS

Dedication vii

Map of Upper Hunter Catchment viii

The River 1

Acknowledgements 235

Further Reading 237

Index 239

DEDICATION

In *The Olive Grove* (Penguin, 2000) I tell of leaving a career in television, moving with my partner, Phillip Adams, to Elmswood, establishing it as a certified biodynamic property, and dealing with the trials and tribulations of planting our small olive grove.

Farms are busy places and we've been lucky to work with many skilled and decent people. Reg Mitchell, our first manager, and Phil Gilbert, who replaced him, have since moved on. The olive grove today owes much to Phil and his son Matthew's hard work and enthusiasm. Yvonne Mitchell is still helping more than ever with management of the house and accounts. Today Gavin Prescott holds the reins when we are away and Colin Watts tends the olives.

There are others: the electricians; mechanics; plumbers; backhoe drivers; trench diggers; engineers; painters; carpenters; builders; pruners; gardeners; fencers; stockmen; shearers; agents; soil scientists; dozer drivers; hay makers; hydrologists; irrigation experts; biodynamic farmers; naturalists; botanists; local historians; nurserymen; fellow beef producers; olive growers and processors and neighbours who have contributed in many ways.

This book is dedicated to all of them and our many friends who have made Elmswood the beloved place it has become.

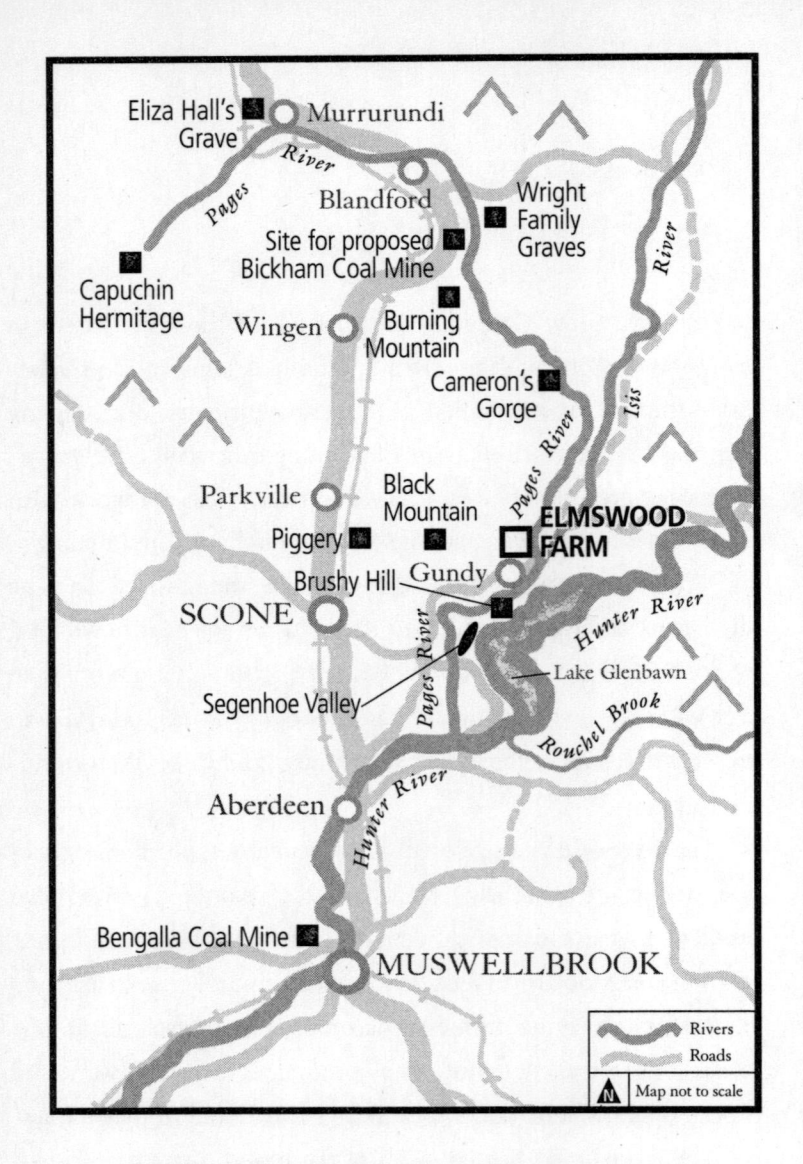

UPPER HUNTER CATCHMENT

1

It's late February 1992. I'm seven months pregnant and standing on the banks of an angry flooding river, feeling incapable of anything except awe. It's rising so fast that no sooner do I find a vantage point than I have to retreat. We're used to floods at Elmswood – they're forever thundering down the valley, but this will be the biggest since the 1950s. It won't subside until it laps the lucerne paddock in front of the homestead, leaving me cut off for days.

This is the day I decide to tell you about our river: the Pages, as in pages of a book. Most of the time it isn't a mighty torrent but a creek you can wade through. It starts with a hint of moisture in an otherwise dry gully about 60 kilometres upstream. From there its flow grows through steep valleys and narrow gorges before arriving in a wide and affluent flood plain. Chaperoned by Australian she-oaks and English willows, it finally loses its identity in the Hunter River, which, in turn, flows into the Pacific Ocean at Newcastle.

For countless thousands of years it was a vital stream for the Wonnarua people. From 1825 they were dispossessed, their river appropriated by the European settlers who came and kept coming, succeeding and failing, trying to make a living and building towns like Murrurundi, Blandford and our little village, Gundy. Old journals suggest that the Murrawin tribe of the Wonnarua, comprising sixteen men, eight women and five children, lived right here. Their ceremonial ground was where the Pages meets the Isis River. I can see it as I write these words.

When the Pages isn't ripping trees from the banks and pounding the bridges, our river supports hundreds of families along its 83-kilometre course. For months on end, sometimes years, it's hardly there at all. At the onset of a drought it begins to disappear, leaving only murky pools and memories. Yet even then, farmers and communities suck from it with their pumps that, on quiet nights, chirrup like cicadas.

Official and unofficial records of the river, of its history and botany, of the life within it, are rare and fragmentary. Even Allan Cunningham's journals of 1823 ignore the botanical details I yearn to know.

These days grown-ups will tell stories from their childhood – of fishing, swimming, diving, boating, camping on its banks, of times when the flow was stronger, the waters deeper, the river healthier. But no journal records where platypus played or kangaroos drank. No poems tell of the turtles breeding, the snakes sunning, the herons, egrets and moorhens seeking food.

Local historical societies refer to the Pages' dramas in their archives, to the people drowned, the bridges smashed and the official openings of the bridges that replaced them. They know the headlines of the Pages, but not the small print. There's little evidence that it's been understood or cherished for its own sake. There's no flow of information about the Pages, no depth of knowledge. The likes of Hans Heysen and Arthur Streeton painted no pictures and the Wonnarua people kept its story in their memories. Some were still hunting and fishing here when Australia was being federated and celebrated. Now their stories are mostly gone.

And it's more than likely the future communities that will suck water from the Pages won't contribute to any continuing tradition or legend. For more and more people are coming and going, buying and selling their houses and farms, whilst the river, except in flood time, diminishes and shrinks.

A community should love its river, and show that love by protecting it. How has the Pages, like so many 'unregulated streams' (for that is its technical description), come to be in the state it finds itself today? Clogged with algae before spring is through, infested with carp at the expense of the native fish, denuded of native vegetation for much of its length and, increasingly, robbed to keep lawns lush, crops growing (especially the lucerne for thoroughbreds and polo ponies) and, ironically, to fill private swimming pools.

On bold bright days I peer into our river where catfish circle their nests, resisting the bullying of the carp, turtles glide like underwater frisbees and eels pose questions with their bodies. And I ask myself: what are the waters telling me? The answer that comes back is: we've taken too much, used it up and never felt the faintest obligation to give anything back. The 178 years of taking has not damaged only the river, it has damaged us as well.

The way we have treated the Pages recently reflects our attitude to the entire world. If the river is to be allowed to flow into the future, we must first discover, and experience, the totality of its beauty and the urgency of its problems.

It's strange. We find energy to wash and clean clothes, so that our children look neat. We hose our cars, so they shine. We drench our plants, so they'll flower. Yet the river, the source of the water that makes our very survival possible, has to struggle on, with little respect or gratitude. If you were to abuse a child the way we abuse the river, you'd be arrested.

I came to the Pages having, as a girl, mucked about in boats on the Murray, picnicking on its banks during trips from Adelaide to Melbourne and fishing with my father. Later, when I leave South Australia to travel around the world, I'll send Mum and Dad postcards from the Thames, the Seine, the Rhine, the Yangtze, the Nile, the Ganges and the Mississippi. Strange that I should end up moving here at the end of the 20th century, sitting on a rock, beside a stream not known to many, counting the turtles swimming in and out of an underwater cave, while my daughter launches herself from a rock. But when she's in the air, before she hits the water, they scuttle away.

What's going on in that cave? Splashed, I'm laughing while Rory (who now insists on being called by her proper name: Aurora), wide-eyed in the water, seeks an answer.

It's September and months will pass before I'll feel it's hot enough to join her. Aurora, on the other hand, as amphibious as most Australian kids, insists summer has already begun. Dragonflies are everywhere, stuck together, copulating as they fly. The ducks relocate themselves downstream, waiting for us to leave.

While it's summer for Aurora, it's spring for the rest of us. The middle of August provides warm afternoons, almond and plum blossom. We re-queen the beehives and take more than 100 kilos of honey, thanks largely to *Eucalyptus melliodora* flowering so profusely. We watch the annual calving across the low paddocks and record a growing number of bird species in the shrubs and trees I've planted around the house. We see little 'torties', as Aurora and Phillip call them, but they're really (mostly) short-necked turtles, clambering from the dams and heading off, so bravely, in all directions. And we take special care in the garden where, in the space of an hour, we see four brown snakes.

Right now, with the changing of the seasonal guard, everyone is busy. Everything is alive.

We live in a beautiful part of the world but, like a marriage that seems good on the outside, functional and cordial, it can be falling apart on the inside. ('I never knew things were that bad,' we say when we hear of the divorce.) It isn't the beauty that reveals the river's essence. Rather it's the extremes of its flow that deepen our understanding.

Every flood is different.

It's past midnight and hasn't stopped raining for two days. While I'm lying in bed dry and warm and comfortable, my mind – like everything outside – is saturated and soggy. I've been worrying for an hour. Worrying that the waters will rise too high, that I'll have to put on my gumboots, climb into the 4WD and wrench the mono-pump out of the river. It's Sunday night and there's just Aurora and me on the farm – one of those rare occasions when there's not a man around. Phillip went back to Sydney anxious to get over the bridge before it disappears; and the farm manager is on leave.

I don't like it. If the waters rise dramatically, I'll have to deal with the second mono-pump on the far side of the property. It's a pump I rarely look at, let alone operate or drag out. Now the rain begins to pelt down until it's so overwhelming that I have to roll out of bed, go downstairs and ensure the patio isn't flooding. But I'm too late – the water is whooshing under the back door into the hallway. I grab towels from the bathroom and make a levee along the carpet's edge, grab the huge rechargeable torch and, ready for action, haul myself into the truck. Hoping that Aurora is still asleep, I drive to the bridge, park beside it, lights on high beam, and watch. I can judge how quickly it's rising by the speed with which it's drowning rocks and branches – and it still has a metre to go before it's over the bridge.

There are two pumps, one on either side of me: on the left, for irrigating the lucerne paddock, a massive Grundfos on its mighty sled, which can only be budged by our biggest tractor; to the right, the little mono, with a permanent cable attached. In an emergency I can hook it on the back of the truck and pull it from the river, pipe, foot valve and all. As the rain increases I manoeuvre the truck along the banks to a position above the mono and hook the cable to the towbar. Then I remember the electricity connection and slide and slip down the slope to the tree where, three metres up its trunk, the plug is nailed. Big bolts have been hammered into the tree to use as a ladder and, rain drenching my face, I climb them and yank at the plug.

That done, I cross our bridge and drive through Gundy, where spotlights shining on the river banks illuminate clusters of people pulling their pumps out too. The other mono-pump, perched higher up the bank, looks comparatively safe. It would take a potent flood to loose it. Or am I kidding myself? I'd shift it if I knew how, but, since I don't, I persuade myself it's safe. Besides, I must get home before the bridge goes under, before Aurora wakes up.

Driving back through the village I see more and more people congregating at the river's edge, torch beams and 4WD headlights fluorescing in the downpour. Moments later I'm back at our bridge – except it isn't there. It's already a metre under water. The river has risen two metres in less than forty minutes.

I panic. What about Aurora? I hope the roar of the river hasn't woken her. Now I have to drive back through Gundy and hope I can make it to the homestead via a back track we don't allow anyone to use when it's raining, because slipping and sliding on it is not only dangerous, but destructive. Even if you don't career into a gully, you add to erosion problems. Yet now I have no choice but to risk it.

I'm wearing old tracksuit pants with questionable elastic around the waist, short gumboots with no socks, and the T-shirt under my Drizabone is already sodden. If I can't get the truck down and up the first precipitous hill, it'll be a wet, uncomfortable, five-kilometre walk home.

I'm driving too fast, the truck flailing around in the mud, and I can't help thinking of the people who've drowned on nights like this. I hear the warnings of the river and the creeks as I prepare to tackle the gully. Bone-dry for a year, it's now filling with run-off. If I didn't know it well, driving across it every day of my life, I wouldn't chance it. On the verge of tears, I grip the wheel, press the accelerator and lurch down the steep incline, in the hope I'll be able to splash through it and slither up the other side. Somehow, slipping and slewing, I make it through and reach high ground.

At this rate, both the mighty Grundfos and the reliable monopumps will be lost or destroyed. And there's absolutely nothing I can do about it. The night, as dark and deep as the river, seems capable of swallowing me; and all I can think about is Aurora, hoping she is warm, safe and asleep, oblivious to my fear . . .

Floods change everything. They drown other memories. Just when you think the river has reached a point of no return, when the scattered pools are at their scummiest, when you haven't seen a fish in months, when even the ducks have given up in disgust, when the last pump has fallen silent, down comes the rain. And down and down and down it comes, until a flood starts a new cycle of life. This is how droughts break in Australia – not with a whimper but a bang.

Then we, the owners, if not the custodians of the surrounding land, take a deep breath, rest a little and wait for the waters to recede. After days of being cut off, the bridge, with its ends washed away, is covered in trees and muddy muck, so we wait for the council to send out the bulldozers to clean up the mess and reconnect the bridge to its banks. Then, when the water is clear, we climb up the tree, reconnect the plug to the power point, throw the switch and hear once again the familiar throb, the heartbeat of pumping. And everyday life begins again.

One persistent story of death – of tragedy – on our river remained elusive until, one evening, the phone rings and Phillip hears an urgent voice pleading, 'Please don't hang up.'

Half an hour later Phillip, very quietly, tells me of the conversation. Or rather of the soliloquy the stranger, Justin, poured from his heart.

In 1971 Elmswood was owned by Justin's father, Ken, whom he remembers with great love and admiration as a giant of a man, good at boxing and horse riding and capable of fixing anything. Then came a night of wild storms and a huge flood. Justin was only eight years old and was ordered to stay home with his mother and younger brother, while his father rode off on his horse into the rain and thunder. Justin's last sight of his dad was as a receding figure illuminated by bolts of lightning.

He never saw him again. Though his memories are confused, he spoke of his father's body being found wedged in a tree when the flood receded – not that he'd been allowed to see anything. 'They took me away from Elmswood next morning and I've never been back.'

Justin told Phillip how the experience had not only changed his life but almost destroyed it; and how every year, as the anniversary of his father's death back in 1971 approaches, he falls ill.

'And now he wants to come back to Elmswood,' says Phillip. 'He believes that if he visits and sees where his father's ashes were buried, apparently very close to the house, he'll be able to deal with it. So, of course, I said yes.'

A few weeks later, Justin, now a farmer himself, arrives with his wife and children. A lot about Elmswood is entirely different from his memories of it. He can't, for instance, find where his mother placed his father's ashes. 'This garden bed wasn't there then, nor was this lawn.' On the other hand, much of the place is exactly as it was, as he confirms with old Box Brownie photographs of his smiling, good-looking father and snapshots of the homestead and farm. One has Justin standing on the now defunct flying fox across the Pages.

For most of the time we leave him alone with his memories. Leave him to walk round the homestead. To stand quietly, sadly, staring into the distance or looking at the ground beneath his feet – hearing the roar of the storm, of the flood, in his head.

Justin is clearly a decent man who, 'protected' from the painful truth as a boy, was denied his chance to mourn. In a couple of terrifying hours, he lost his father, his home and the life he'd known up to that fateful night.

When we say goodbye, we invite him to return any time he likes. As the weeks pass, we wonder whether Justin found any peace in coming back to Elmswood or whether the experience only intensified the pain. Later we hear that his visit has helped to heal his grief.

Justin hasn't returned. When I try to contact him, a few years later, I find he's sold the farm and moved to another. I track him down and he confirms that the past is now the past, that Elmswood's greatest tragedy and this sad story of the Pages is now what it should be: a memory, a significant memory, but no more than that and not an event to be lived and re-lived for the rest of his life.

Now Justin's story is a part of our story and we, too, remember it on nights of storms and flood.

2

Given a filigree of fine blue lines on the map, it's hard to deduce where the Pages River begins. There's no 'spot' to officially designate it, though the Geographical Names Board says it rises 11 kilometres west-south-west of Murrurundi. So they must have measured it to an exact point, but where that is I don't know. Neither does anyone else I speak to.

I follow the pale dirt track called Pages Creek Road as it winds alongside the river, passing beneath massive, eroded sandstone outcrops, keeping company with the main north railway line. The track sweeps through the hills on its way to, or from, Nowlands Gap in the Great Dividing Range. Alongside the strength of the rails, sleepers and steel bridges, the river looks frail and puny. Mature she-oaks lean and arch and knock each other over, vying for position. The old trees are almost on their own now, with only a few youngsters amongst them. Most of the eucalypts on the lower slopes have been cleared by farmers. The hills, too steep for cattle, give them some protection. Here the plants become more diverse.

Cattle rest in their straggly shade, or stand in the river up to their hocks, muddying the pools. Basalt rocks, deep grey, grow in size the higher you look, seeming to reverse the laws of perspective. Up above they balloon into boulders, creating crevices for delicate waterfalls when it rains.

At the end of the road there's a property called High Valley. Its lowest point is 600 metres above sea level and it peaks at twice that height. Here a hermitage, a place for prayer and contemplation is being created by the Capuchin Franciscan order, fondly called the 'Cappuccino Monks' by the locals.

There's a yarn that hundreds of years ago the Viennese concocted the cappuccino after the Turks monopolised the coffee-bean market, inflating prices until they were prohibitive. This led to a milky coffee, served with a dash of chocolate on top, evoking the colours of the Capuchins' simple costume. While Brother Henry tells me the story with obvious pleasure, he concedes it may be apocryphal.

(Coffee expert and historian Ian Bersten assures me it is. Cappuccino as we know it today was probably first made around 1906 in Milan. Capuchin friars in 16th-century France were known to be charitable and they offered a drink made from chicory root to the poor if ever they came knocking. That drink was possibly called a 'capuchin'. Bersten believes coffee was added later and, when frothy milk came along, 'it was simply an extension to call it cappuccino.')

In 1998 the Capuchin order bought the 640-hectare cattle property, intending to create a retreat beside the original wooden cottage. 'Silence is the faithful guardian of the inner spirit,' says the text in their brochure. Given that the map is ambiguous, I've decided that the Pages, my Pages, begins here – because there's something deeply comforting about having a Catholic hermitage at the head of our river. Blessing it. Creating 'holy water', no less. I am, after all, a lapsed Catholic with fond memories of retreats where groups of us, uninterested in the 'inner spirit', puffed fags in the bushes in preference to discussing God.

The approach to High Valley is through a landscape on familiar terms with aridity. And it was this drive, almost an hour from Elmswood, that made the hermitage seem more dramatic and the faithful guardian of what the country was like in earlier times. No introduced stock has been allowed to graze here since 1998, though it's been impossible to exclude the odd rogue steer, lamb or goat. Already there are clear signs of recovery. This land is reserved for the natives: the wombats, wallaroos, kangaroos, wallabies, koalas, numerous lizards and snakes and a growing inventory of birds.

The difference is profound. There are many grasses, both introduced and native, clovers and creepers galore, and I'm confronted with the reality of the damage hard-hoofed animals do. Yet the response to High Valley isn't always approving. While some see it as a masterpiece, others deplore a waste of productive land and a provocation to bushfires. I'm on the side of the friars.

Brother Henry is waiting, wearing his cappuccino costume, outside the wooden cottage he shares with Brothers Paul and Albin. It will take a couple of days to explore the place fully, but this trip has one purpose – to see a waterfall I've heard about. We hop into Henry's 4WD and head off along the new roads they've cut into this black basalt valley. The principal reason for the roads is to give access to the Stations of the Cross the brothers have erected from treated pine that, instead of leading to Golgotha, take us to the place where a new hermitage will be built.

'Why here?' I ask as we inch our way up the roads of black earth, which looks and smells edible.

'It's remote, beautiful, cheap and close to public transport,' says Henry proudly, 'which makes it an ideal place for prayer. Don't you feel you're in Heaven?' And, yes, we are surrounded by a spectacular panorama. We glimpse a blue shimmer beyond the trees that turns out to be smoke – where Brother Paul is burning stumps and old blackberries that have been cleared from the side of the tracks. And there's another blue shimmer – of a lake, created years before, now surrounded by willows and filled with lilies.

We park the truck and I follow Henry along a narrow path – my small Blundstones stepping in the impressions left by his huge sandals. Paul has preceded us, in recent weeks, marking the route with a mattock. We pass his tools, left resting on a big rock. He's been working his way around the property, not so much making formal tracks as quiet suggestions. Here he's lifted a few rocks. There he's flattened out a couple of metres. Sometimes he's placed markers and posts to guide the pilgrims and he's made little wooden bridges across tiny gullies – just enough to human-ise the wilds, to transform bush into a rugged garden.

We are on our way to what Henry has officially named High Valley Falls. And despite the antiquity of his order, he's a very modern friar who makes maps on his laptop, on which he marks the hints of pathways, the Stations of the Cross and the other features he's had officially named – or rather christened.

There's Mount MacKillop, 1190 metres high, named after Mary, who established the St Joseph order in the South Australia of my childhood. There's Wonnarua Ridge and Kamilaroi Plateau, named after the all-but-forgotten Aboriginal tribes. There's Mount St Francis at 1134 metres, Ben Hall at 1220 metres – named for the bushranger who, born in Murrurundi, became almost as famous as Ned Kelly. Ben's peak is flanked by the arresting presence of The Constable at 1190 metres. There's also Joseph's Rest, and Mount Torreggiani, in honour of the second bishop of Armidale, who was a Capuchin. But Henry's most creative idea was a Mount Sinai, in honour of a neighbour landowner called Moses. Unfortunately Mr and Mrs Moses were unamused and the christening was cancelled.

It's soft underfoot as we climb towards the falls. Both Henry and I are enthusiastic bird watchers as well as amateur botanists; we compete to point out the finches and bush wrens – and delight in the cacophony of flirting parrots that seem to escort us. Around our feet is *Hibbertia scandens*, the yellow flowering creeper featured on a recent 45-cent stamp.

The track steepens, the trees grow taller and thicker, and we push through creepers until we come upon the surprise of a picnic table that Paul and Henry have placed under an ancient 40-metre angophora in front of the black rock, where – despite the dry – water is cascading down. Still higher above us are she-oaks and the spears of grass trees. But even here isn't close, so Henry hoists his skirt and sherpas me over rocks until we stand at the very edge of the falls. Where, feeling appropriately religious, I kneel and cup my hands to take a sip of the cold, pristine water. To walk to the top of the falls and back around the Kamilaroi Plateau we'd need another day.

White occupation from the 1820s saw this valley hacked into large estates and flogged for logging. There's evidence of it everywhere: vast stumps, metres high, still clearly displaying axe marks, and felled logs that didn't make it to the mills. Nonetheless, because cutting down trees was such hard and dangerous work, the process was selective. Not like the clear felling employed today where forests are simply razed. Henry is excited by the prolific growth of the understorey since the eviction of sheep and cattle. We pass hundreds of bottlebrushes, wattles, cycads, ferns and creepers and rush to identify them. But there are still unwelcome invaders to deal with – principally blackberries. An unforgiving Brother Paul has been dealing with these sinners. (Paul joined the order in 1972, and would have been a geologist if he hadn't heard the call. He set himself the task of researching one of the local curses, preparing a hundred-page report: 'A Study on the Control of Wild Blackberries.' Now he's under orders to report to the Vatican, where he'll spend the next few years in that extraordinary library, deciphering old Capuchin documents.)

Surrounding High Valley are cattle and goat properties. The goats were introduced to eat the blackberries and can be coerced into doing a pretty good job. The hills have been fenced with netting and electric wires to keep the goats focused on their task – but the same fences are disasters for the native animals. Opening one gate I find an echidna caught between a trip wire and the netting. In the past, I've been impressed with the goats' efforts, marvelling at how weed-free their paddocks have become. In any case, where you see steep hills grazed until they look like suburban lawns, you wonder whether goats are such a bright idea. They've made much of the area productive again, but at the cost of diminished biodiversity.

Brother Paul's alternative to goats, I'm alarmed to learn, is chemicals. He's already spread many tonnes in the last five years – and that's treated just a fifth of the hermitage's unwanted crop. Somehow it doesn't seem to fit with the Stations of the Cross, particularly when the Pages begins its life here and provides the township of Murrurundi with its drinking water. But something has to be done, otherwise blackberries will consume everything in sight.

An 1828 map names the Pages the McIntyre, and the Isis as the Pages. Others call it the Page River or the Page's River. One map has it end at the confluence of the Isis, with the Isis continuing on to the Hunter, and looking at a map that would appear to be the case. The Pages does flow at right angles into the Isis, so why continue to call it the Pages from that point? And why the Pages in the first place? There were no surveyors called Page and no person called Page living in the Hunter Valley in the census of 1928. Records reveal that, when Henry Dangar was conducting private survey work for Potter McQueen, he traced a river already known as the Page's. The only hint of an origin comes from Potter McQueen's marriage to Anne Astley, whose sister had married into a family called Page. Is that the association? Or just a coincidence?

I drive across the Pages at a ford adjacent to the Murrurundi footbridge. Although it's the first time I've used it, it's thoroughly familiar. A photo of it appeared on the local telephone directory with Chloe, Aurora's friend from school, splashing through the water. Both Chloe and the Pages look gloriously healthy.

I park outside a 19th-century brick building on the New England Highway that announces itself as 'The Literary Institute'. It's now headquarters to the local historical society and the domain of Barbara Riddell. Barbara likes to call herself a blow-in, though she moved to Murrurundi thirteen years ago and has been adding to the archives ever since. She's set herself the task of computerising all the newspaper clippings, documents and photographs, and today has something special for me. She's discovered a photographic image of the Pages River printed on one of those souvenir tin trays you used to see in every kitchen.

It's an image of the place I've just visited – except that today the falls were a trickle compared to the torrent in the photo.

Most images from the 1860s reveal the district all but denuded of native trees. Today, along the river through town, there are willow, claret ash, cottonwood, broadleaf privet and the occasional she-oak; and in the main street there are bunyas, holm oaks, jacarandas, English elms and a spectacular *Grevillea robusta*. The introduced trees are what make the town so colourful every autumn. But gone are many of the yellow box, she-oaks, angophoras and ironbarks along the valley floor.

'The river flow of the Pages is as whimsical as a yoyo,' Barbara says, and explains that there's one historical fact she's sure of. It's that settlers have been complaining of the Pages ever since they arrived. 'You're not the first, Patrice. People have been grizzling about the lack of water in the poor old Pages for 150 years,' she adds, referring to articles published in the *Maitland Mercury* in the mid 19th century.

If it wasn't too empty, it was far too full. Thus in one letter to the editor, dated 1857, a man curses the flood that caused him to miss his wedding. By the time the waters had fallen, his bride-to-be had jilted him and run off with another chap.

> Gentlemen, I am a forgotten swaine, and I have to thank the listless, spiritless, apathetic put-up-with-anything denizens of the Page for it. The townships of Murrurundi and Haydonton are separated by the River Page, which when swollen comes down in a perfect avalanche, continuing impassable for days and sometimes weeks.

In 1858, reports of the same crossing have the River Page not running at all.

Arms piled with documents, CD-ROMs, photos and catalogues, I leave, unable to thank Barbara enough. 'The Riddell family motto is to share,' she says, beaming. She's still waving in the doorway as I drive off.

3

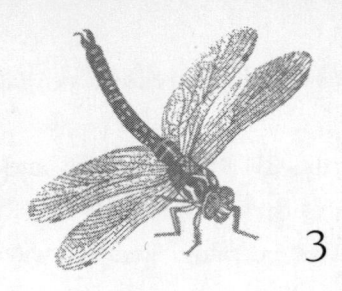

Where do all the floodwaters go? The question is something of a joke around here. What worries some, during floods, is the thought of so much water vomiting into the Pacific Ocean and heading for New Zealand, as if the Pacific or New Zealand needed it. Whereas back in Gundy, back in the Upper Hunter Valley, we always need it. There's a deep desire within all of us to catch it. Every drop. We use technical terms like 'water harvest', 'water cropping', 'diversions.' But what we're really after is to capture the Pages and keep it for ourselves. Just as we muster sheep and cattle into yards, we want to muster water into dams.

At an early meeting to discuss how the river should be properly protected and appropriately used, a man whose family has dominated the district since the early 19th century, totally ignorant of ecological arguments, said it was a crime to let the flow go down to Newcastle. 'We should dam the river up,' he said, defiant and angry. I wanted to tell him horror stories of the Aswan Dam on the Nile, of the environmental ruin caused by similar dams in India, China and Indo-China. But what's the point?

In the meantime, there has been some harvest. Every dam, on either side of the Pages, contains run-off from the rain. And the tributaries (or rather contributories) that lead to the Pages – creeks like Splitters, Scotts, Warlands and Kewell – all retain some of the rain that falls in the hills for flora, fauna and stock. However, there's no big municipal dam to hold the floodwaters back, to act as an oasis for tourism, making the Pages a 'regulated river' like its big sister, the Hunter, which flows into the giant Glenbawn Dam nearby. There have been propositions to build such a dam on the Pages over the years but, thank heavens, such ventures are increasingly out of favour. All over Australia, all over the world, they've proven to be a cure worse than the disease.

The best achievement a landowner can claim is to 'capture' the water in the soil itself. Where it should be. Where it used to be before white settlement and the destruction of the native vegetation. There, living in the soil, water can instantly be put to use.

This is a central ambition of biodynamic agriculture. Each area of land has its own desire and it's not always evident what that is; consequently, paddocks respond differently to my attention. A responsible farmer's job is like that of an art restorer: to reveal the full beauty of the original and to let the painting, or the paddock, breathe again. It can be as revelatory as the Sistine Chapel's restoration, which enables viewers to see, for the first time in centuries, the astonishing vibrancy of the frescoes. That can happen with paddocks, with hillsides, with valleys. If you get it right, you can make them shine, and then they can retain at least some of their beauty during the toughest drought.

On a good day it's impossible to believe the Pages is anything less than perfect. I have, surrounding me, a landscape that can break hearts, cause us to weep and conductors to break their batons in the knowledge that their art cannot compare with life. But the landscape around me could be far, far better than it is. All you need to do is honour it, listen to it, help it.

Despite some water being held back in the five dams on the Hunter River and the thousands of little dams along the tributaries, most of the local water ends up passing through the estuary into Newcastle harbour – and sometimes it fills the harbour with the willow trees and she-oaks wrenched from the banks. Once the water leaves Gundy and joins the Hunter River above Aberdeen, it flows through Muswellbrook, Singleton, Branxton, Maitland, Morpeth and the Novocastrians' urban sprawl, where, as long ago as 1840, the municipality worried about sediment clogging the harbour.

It was from Newcastle, the second port established in New South Wales, that explorers and settlers ventured into the Hunter Valley. Within twenty years all the best land had been claimed. Reports from explorers Allan Cunningham and Henry Dangar described dense vegetation along the Hunter and its tributaries. Settlers like Peter Cunningham and Henry Dumaresq wrote about rich river flats. Immediately after settlement in the 1820s, however, due to the value of the timber so easily accessible along the river banks, things began to change.

The clearing began with a vengeance. It had been dense rainforest, although it was commonly called 'cedar brush'. The flora included giant red cedar, figs and myrtle. Right on the water's edge tall, smooth-trunked eucalypts vied for position with she-oaks. Most of the cedar brush was taken by the 1830s and it would be years before small pockets would be found hiding in the Upper Hunter and at the head of the Pages.

It was fortuitous for the settlers that there were trees to fell and sell, for agriculture had a difficult start in the valley. No sooner was the land cleared and crops sown than the first drought was recorded. By the end of 1826 crop after crop had failed, and the weather wouldn't improve until late 1829. The river flats, abundant with economic hope, turned into sand; the hillsides, now devoid of trees, were eroded; and, in dry weather, clouds of dust told the location of the flocks of sheep, which doubled in number every two to three years. Soon the sheep were driven north over the Liverpool Ranges in a desperate search for feed, leaving trails incised in the hills that to this day can be seen at Elmswood.

When it did rain, it was clear that the settlers' rape and pillage had changed the valley forever. In particular, river-bank erosion was horrific. In March 1832, just two years after the breaking of the drought, floodwaters gushed to the mouth of the Hunter, where Alexander Livingston, Newcastle's Harbour Master, feared his port would soon be closed by silt. The changing channels and shoals at the entrance made navigation so difficult it was becoming unsafe for seamen. And this was after just twelve years of white settlement in the Hunter – and six years at the head of the Pages.

Thanks should go to Mr E.W. Moriarty, who prepared the first report on the floods of the Hunter in 1868 – and to the enlightened minds of 1870 who suggested a Royal Commission be held to investigate the Hunter and its habit of flooding. For, by this stage, floods were so frequent that landowners were counting the acres of lost land – from acreages only recently acquired. William Dangar complained of losing eight particularly valuable acres of river flats after the 1867 flood alone. Trees along the banks had been logged and those left standing fell into the floodwater, unable to withstand the force of the flow. At Singleton, the Hunter had almost doubled its width.

The Royal Commission defined floods according to their magnitude and named those of 1820, 1857, 1867 and 1870 as being 'firsts' or 'big', with sufficient power to terrorise everyone living along the rivers, where water levels would rise fast and far above the gauge heights.

Things kept getting worse, and the erosion was now so immense that everyone was echoing Dangar's protests about being robbed of river flats. Furthermore, the course of the river was changing dramatically from flood to flood. Something had to be done. Nonetheless, the Royal Commission refused to recommend that floods be shut out of various areas, rightly worried that interference might make matters worse. Also, in the era before the bulldozer, carrying out erosion repairs was all but impossible.

Despite the fact that 90 per cent of the silt was still going out to sea, Newcastle harbour, receiving its bounty of Hunter topsoil, was becoming less and less navigable. The amount of soil being lost was staggering. Between 1915 and 1987 approximately 4048 hectares of productive land was washed into the Pacific. Some of it came from the Pages River, especially at Segenhoe Valley, these days the principal home of horse studs, and lucerne and crop production. Between 1946 and 1995, many hectares disappeared from these alluvial flats.

No one knew how to tackle the problem of erosion, but at least there was the technology to dredge the harbour. This began in 1859, with 40 000 barge tons of silt removed. Twenty years later they were taking 2 million tons a year out and, by 1952, the dredging took 5.4 million tons. For a century, dredging was a normal and necessary part of harbour management.

Reports, reports, reports. In 1877, 1880, 1894, 1897, 1899, 1901, 1903 and 1913 more experts conducted studies, more bureaucrats toiled over documents. So many people were worrying about the Hunter – but not really on behalf of the river. Their concerns lay with the property owners and their profits. With little understanding as to what constituted the 'health' of a river, virtually the entire focus was on economics. It would take another half century and another investigation, the 1948 *Report of the Hunter River Flood Mitigation Committee*, commonly referred to as the Huddleston Report, before substantial work got under way to repair the river and manage the eroding floods.

While it's clear a lot of effort was directed at trying to understand the problem and work out ways of curbing the effects of flood, it wasn't until 1950 that a new concept was put in place. The Hunter Valley Conservation Trust wasn't just another organisation to investigate flood damage. The Trust, the only one of its kind in Australia, had the power to raise revenue from everyone in the catchment area and to use the money to carry out flood mitigation. There was no shortage of ideas about the best way to tackle this immense problem. Given limited funds, there was also the question of priorities. It would take the greatest flood of all, in 1955, to get things moving. The flood of '55 is still spoken of as one of the great disasters in living memory. Not only did it cause loss of life and destroy previous river works, it also seemed to demonstrate that attempts at mitigation can actually make matters worse.

So as the floods receded, a new organisation came into being: the Hunter Valley Research Foundation, which, these days, would be described as a 'think tank'. Its task was to think about 'taming the Hunter Valley'. And the thinking led to change throughout the catchment with beautiful, poetic willow trees being planted – now considered a highly contentious decision – to hold back the erosion. Looking at the graphs of the dredging figures, it's clear that the biggest beneficiary of the willows has been the harbour. Within twenty years the need for dredging dropped 50 per cent, but that didn't mean erosion had stopped. A survey commissioned by the Trust in 1987 said that if all the gullies and eroded streams of the Hunter were laid out end-to-end, they'd stretch for 10 000 kilometres. In other words, 81 per cent of the rivers were still suffering from erosion. On average, rivers were between two and four times wider than before white settlement. They might have looked pretty, but their breadth was symptomatic of a crisis.

These days willows (except weeping ones, *Salix babylonica*) are deemed a curse in our catchment. However, you'd be hard-pressed to find an old-timer with a bad word to say about them – nor would anyone who lived through the 1955 flood. Willows gave the Hunter a chance to hold back the silt, and stabilise some banks. No other tree could possibly have done it. Their easy propagation, swift growth and undemanding natures have seen them flourish on every stream in the Hunter Valley. Years later, when they begin to age and break off and float downstream, they create a problem. Even their leaf litter in autumn is condemned. What rivers need, we know now, is carbon – preferably from hardwood. Hardwood that sinks and lives another life underwater will ensure river health, with the help of river creatures big and small. Hard, heavy wood that takes centuries to rot is what is needed.

That's one of the big problems with willows. They don't sink. But, as long as the 'approved' weeping willows live, they remain lyrical and lovely and won't cross-pollinate like the upright varieties that quickly spread. You can't stroll up to any native tree, break a piece off and shove it into bare earth and know it has a good chance of surviving, thriving.

Yes, it would be better to see only native trees along the river banks, to recover some of the lost look of the riverscapes before the 1820s. But, similar to Aurora with her memories of *The Wind in the Willows,* I feel a great affection for the trees, and especially in August with their pale, soft green blossom. There's an ancient willow in Gundy thriving in a gully a few metres from the Pages.

Because it's so protected it's achieved a perfect form, reaching high, with a few horizontal branches supporting its weeping cascade. Earlier this year, I cut off two bits, and stuck them in our garden for posterity.

4

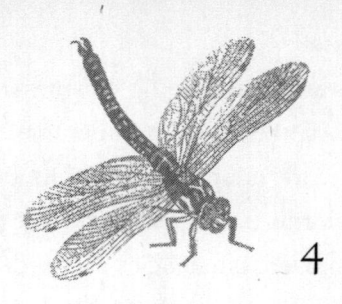

Just as we're about to start working out our water entitlements, we hear there are plans to gouge out a coalmine south of Murrurundi. I ignore, even scoff at, the suggestion. A coalmine in that glorious valley? Across our river? It could never happen. No one would let it happen. I'm too busy with real issues to worry about something so hypothetical, so wildly improbable.

The lead-up to our centenary of Federation was, inevitably, a time for soul-searching – and the national debate focused on the two R's: reconciliation and the republic. In fact, there was a third R of comparable importance: R for rivers. Many of us thought it was time to apologise to our river system too.

Numerous government departments, scientists, researchers, Landcare, Bushcare and Rivercare groups – as well as farmers – were working out plans to make improvements. Clearly 'things' were about to change.

Had environmentalists finally been heard? Had the bureaucracies experienced collective epiphanies? The principal reason behind this attempt at a 'new way of doing business' came straight from the Council of Australian Governments (COAG) National Water Reform Framework, adopted in 1994. Its aim was to implement market-based reforms that reflected the true cost of water. Subsidies would be removed, efficiencies demanded and new projects would have to demonstrate ecological sustainability as well as economic viability.

It wasn't the people's idea. Nor could the New South Wales Department of Land and Water Conservation (the all-powerful DLWC), nor the ministers of the Departments of Agriculture or the Environment, and certainly not irrigators, claim the credit. The idea came from a group often pejoratively described in rural areas as the 'Shiny Pants Brigade' – faceless civil servants who spend their lives rewriting the rule books, and they had turned their attention to water.

The COAG agreement set the wheels in motion by making water a property right. Once it was decided 'who owns what amount to use when' the owners could trade that right in the open market. Water was no longer going to be linked to a specific piece of land. Like everything else in this modern world, water would undergo a process of commodification. Economists would now have something else to include in their data and theorising – and entrepreneurs a new product to sell. And local real-estate agents would have something else to advertise in their windows.

That was just one of the revolutionary propositions in the plan. The other was bigger, bolder and more beautiful. Rivers would, by law, be entitled to some of their own water for their own sake. Instead of allocating river water exclusively to all and sundry, the river would receive its own allocation – and that amount, that percentage, would be at the heart of hundreds of heated discussions across rural Australia in the following years.

On hearing this news, Ken, a manager at a neighbouring horse stud, and consequently one of the biggest water users in the Segenhoe Valley, called a meeting to see if we could set up a local association to help us conduct the discussion and deal with the immense amount of data being dropped into our letterboxes. A few weeks later, the Pages and Tributaries Water-Users Association was officially established – and at the head of the agenda were the new rules and regulations being posted by the DLWC.

This state bureaucracy with 2500 staff in ninety-nine offices in New South Wales towns subsumed organisations like Soil Conservation and Water Resources, bringing the catchment area into the same bureaucracy as the rivers. This made all sorts of sense and there were high hopes for the process, if some fears about bureaucratic size.

The DLWC was facing an immense and unprecedented task: to restructure the management of, and our attitude to, every aspect of our river systems. Let it be said that in its early incarnation, the department had been responsible for many catastrophic mistakes – including immensely wasteful irrigation arrangements and the reckless granting of licences. And we're not talking ancient history. For instance, the Riverina was still fast-tracking developments in the 1970s, no matter how absurd the project.

Hence the proposal to grow rice in Deniliquin, where hundreds of farms would use 1000 megalitres of water a year to do so. By the mid 1970s some projects were proving disastrous. The ground watertable had risen markedly and salt began to be seen everywhere.

At the same time licences were cheerfully granted to anyone who wanted to pump from the Pages – until the mid 1990s, when they slammed on the brakes. But the moratorium had come too late. Every river in New South Wales was over-allocated, often by orders of magnitude. If every irrigation licence holder switched on their pumps simultaneously, every river would run dry.

Now the DLWC had to get itself, and the rest of us, out of this mess. And were we, the landholders and licence holders, willing to help? Imagine the Department of Lands deciding that all the quarter-acre blocks in Australian suburbs were too big and announcing a decision to take 10 square metres from each and every block. That's how people felt about their water allocation.

'Oops, sorry, you've got a bit too much water there. We've got to take some back.'

'Why?'

'Because we're going to give it to the river.'

Little wonder that even before the first meetings were held people were getting very, very angry.

We were advised to view the Pages catchment 'as a whole' and to welcome every water user to our association. Not that the issue was entirely foreign to locals who had, for years, been forming unofficial groups. For example, the irrigators on the Isis and the upper Pages had been holding meetings but, unfortunately, no one had thought to keep minutes or records of agreements. It was all anecdotal, based on memories of who'd said what and handshake agreements. 'Harold needed to finish his crops, so I said, "Yes, old son, go for it."'

But when, during a previous drought, I had spoken to mothers at Aurora's day-care group, I was all but deafened by complaints about various men (husbands of women not present) pumping too much water to the detriment of others. There was a lack of community consideration and bitter criticism of irrigators' selfishness. So much for gentlemen's agreements.

At our first meeting we elect a President, a Vice President, a Secretary and Treasurer, and open a bank account. The joining fee is set at $20. Even at this discount price, we don't achieve a 'whole' river group because no one from the upper Pages comes to the meeting – not even a representative from the Murrurundi Shire, which relies on their allocation for town supply. So we're not off to a flying start. Maintaining a bad tradition in the district, the association is representative of only half the river – beginning at Cameron's Gorge, a designated nature reserve that creates a north-of or south-of mentality. (Water-users from the Isis River, a large tributary of the Pages, become members, but they vote to manage their affairs separately.) So before it's begun, there's a fracture. Sad that a river only 83 kilometres long can't muster enough members to form an appropriate association.

And it wasn't for lack of trying either. A few of us made phone calls to those upstream trying to entice them along, only to be met with comments like: 'I moved up here so that I could get away from all that rot' or 'When there's no water I can't irrigate anyway'. One of the refusals was openly aggressive. 'I don't want to go to meetings with those gorillas down in Segenhoe Valley.' In due course, the DLWC would state that the Pages River Management Plan would include the whole river. End of story. Not negotiable. But even that made no difference.

How can an idea of the magnitude of river management be tackled? Well, we dealt with an even bigger concept in the past. During the 1890s the idea of Federation was debated from coast to coast in scores of legislatures and town halls until, against all the odds, a nation was brought into being. Perhaps the idea of saving a river and negotiating its use isn't as glamorous or as finite, because we're dealing with an ongoing process, not a one-off event. The proposed reforms would not only require us to reassess our rights, but to accept the rights of nature. The task would involve our feelings, our philosophies, our political and economic, even our religious, views.

History has taught us that when humans are *really committed* anything is possible. Consider the Manhattan Project. When faced with the thought of Hitler having an A-bomb, a group of highly motivated physicists gathered in the American desert to head him off. Perhaps we need people to feel that threatened, that imperilled, before we can persuade them to change 'things'. As the months and years pass, I learn that we are at war on every New South Wales river. Trouble is, we're the enemy.

I arrive early for the next meeting at the Gundy Memorial Hall and Garry Hunt, the DLWC's Executive Officer for Water Reform, is waiting outside, his bulging briefcase propped against the railings. Garry, a civil engineer, 'builders and destroyers of the environment' he says with a bleak laugh, has been granted the unenviable task of selling the reforms and negotiating the implementation throughout the Hunter. He's been with the department during its various incarnations since 1975, working on the Murrumbidgee Irrigation Scheme and the Hawkesbury–Nepean Catchment Management Trust. He's also the DLWC Executive Officer of the Hunter River Management Committee, which advises the Minister on all reforms.

'The first thing you have to do,' says Garry to his small audience, 'is work out how much water you actually take now and when you take it.' Everybody in the hall looks at everyone else. It's a simple question, yet impossible to answer. Nobody's been measuring or keeping records and you can't help feeling that some of us are reluctant to make disclosures. We vote to set up smaller groups to find the facts, and undertake to meet again in a few weeks' time.

'It's a huge social experiment,' says Garry as we leave, and I feel like a lab rat.

We're in the era of 'public consultation' though many of us who are 'consulted' consider the process to be, at best, window dressing and, at worst, a monstrous deception. In many areas of public activity, conscientious community members, the volunteers who spend hours doing good works, often find that crucial decisions have already been made. The modern process of consultation can make the activities of the old Shiny Pants Brigade look democratic.

While the New South Wales Minister for Land and Water Conservation has the DLWC to carry out his instructions, these reforms need more staff and a whole new network. The Minister will call on the community at large to help with the reforms but, in the meantime, appoints seven people to head seven committees assessing the seven catchments in the state of New South Wales. Dr Wej Paradice has been chosen to head ours. When I first hear about Wej I ask people to repeat the name. But Wej it is, an acronym for William Edward John.

What sort of person would accept such an appointment? Who would want it? Wej was an inspired choice – and not just because he was born in Scone. One of the first graduates from the University of New England's Natural Resource Management program back in 1974, Wej then worked for CSR in Sydney before heading off to Colorado and Idaho where he fell in love, married and completed his PhD, with Water Management at the core of his studies. Back home in Scone he spent months helping his dad during the 1980s drought before being appointed Director of the Hunter Valley Research Foundation, the private organisation set up after the 1955 floods to see what could be done to mitigate the Hunter's misbehaviour. As well, he's a Trustee of the Hunter Catchment Management Trust. Now, if all these organisations sound confusing, that's because they are. Everywhere I turn I find another organisation, another self-appointed group, another cluster of researchers.

Despite a full calendar, when the call came to head the Hunter River Management Committee, Wej still said yes, adding wryly, 'You don't do it for the glory.'

I'm still on the phone when Phillip and Aurora return from chipping Bathurst burrs. It's absurd to attribute human characteristics to weeds but, on Elmswood, we regard burrs as evil. They start off small, lurking beneath rocks or hiding amongst bigger weeds or under fallen logs. Then, suddenly, they explode like fireworks, and their spikes are lethal. Get pricked a few times and your flesh begins to protest with swelling fingers and painful hands. Aurora and Phillip are a double-act when dealing with them – she can spot a Bathurst burr wherever it's hiding and he'll thump at it with a long-handled hoe or, if it's too big, prune it with secateurs. Working together, they can kill thousands in a session.

Bathurst burrs make Phillip angry and he's returned smouldering with rage. He lingers by me shuffling through mail, willing me, in the silent way only men can, to hang up, put on the kettle and make him a cup of tea.

Succumbing, I say, 'I'll call you back.'

And at this moment Phillip asks, 'When did the tree by the pump fall over?'

'What tree?'

'The solitary she-oak. The big one by the Grundfos.'

'What!'

'It's horizontal. Lying across the river. From one side to the other.'

I can hardly contain myself. Phillip has to give up the idea of tea and drive me down to view the latest problem. We stand on the bridge and I can't believe what I'm seeing. If ever there was a wanted, needed, necessary tree it was this one. One by one, over the years, we've watched she-oaks collapse into floodwaters and head off downstream, but not this one. It had remained defiant – only to topple over on this windless day when there is hardly enough water in the river to fill Phillip's kettle.

It hadn't looked like it was in trouble, though part of its root system was undermined by the last flood. It had stood tall and straight over a crucial waterhole where the foot valve of the Grundfos dangled. The tree seemed to protect the hole where carp circled and, from time to time, platypuses played. The tree was lying at right angles to the river, in perfect alignment to the bridge on which we stood, and in falling it had dragged a lot of the river bank with it.

It's interesting how a solitary tree can provide the illusion of a gentle, almost landscaped stretch of river. Now the bank was naked and seemed ashamed of itself. Not only would I have to look at this raw ugliness every day, at the cavity left by the tree, but I'd feel guilty about it. How could I have expected a lone tree to take the full violence of the river and survive? It had done its best but, exhausted from the task, had died.

How to make amends? How to get it right? I know how high the floodwaters come up these days, supposedly higher than ever and, according to hydrologists, who measure such things, with increasing speed. Why? Precisely because there are fewer trees on the banks to buffer the floods and slow the flow. So here's yet another job to do, another problem with which to contend. We go home and Phillip finally gets his pot of strong tea.

'Did you see the tree, Mum?' says Aurora, returning from the veggie garden with some late-in-the-season asparagus. 'Unreal!'

Later, alone, I return to the bridge and stare at the tree, knowing that it will push up against the bridge during the next flood, increasing the possibility that it could be washed away. I half like that idea. The bridge being demolished and leaving us stranded, cut off. It's an ugly bridge anyway – all brutal cement, with no attempt to make it elegant, let alone attractive.

For years there'd been a pretty little suspension bridge made of cables and planks, where pedestrians felt deliciously precarious, and below it was a ford for vehicles. The ghost of the suspension bridge is still there – the great wooden posts that held it aloft and the broken cables rusting on the banks. A flood decades earlier eventually tore it down and twenty years later a new cement bridge was erected. Now it's far too short and doesn't reach the eroded banks. The bulldozers have to shove huge rocks into place and pile gravel over them to return it to active duty.

The poor tree will have to be removed. But how? I phone the DLWC and tell Ken what's happened.

'You have to get permission to move anything in the river,' he says quickly, in case I had any illegal ideas. Which I didn't.

'Ken, it'll have to be moved. It's just before the bridge. It blocks the river.'

'Chainsaw,' says Ken.

The image of Phillip in the river being careless with a chainsaw flashes before me.

'We're not doing it,' I protest, 'you're the experts!' I hang up and stand by the fax machine and, in due course, a six-page form emerges that, when properly filled out, will give me permission to touch the tree. Little did I realise the tree would loom larger in its death than it ever had when protecting the fish-filled waterhole, its boughs alive with honking ibis.

Weeks later, we're back at the Gundy Memorial Hall, coming to terms with water reform.

'Basically what's different,' says Garry, 'is that the environment has a right to some of the water.'

So we divide the river into sections. I become the contact person for the area between Cameron's Gorge and Gundy – a stretch that has only a few irrigators, many of them inactive, but where there's increasing demand from housing and stock and domestic water use.

'We're in Stage One of the changes now,' says Garry. 'All water licences on unregulated streams like the Pages will be converted from hectares to volumetric measurement.' He points to a graph in one of the booklets we've received. 'How your licences are converted depends on what you use the water for now.'

How I use the water now? This seems a silly idea. What if Elmswood wants to do something different in the future? To switch from cattle to cropping? To plant wheat as well as olives? What if we're in the middle of a big restructure?

Garry reads from a list: 'If you water perennial pasture and you're not a dairy, one hectare converts to 5.5 megalitres, but dairies convert to 7.5 and lucerne 6.5 megalitres. Turf will get you 11.5 megalitres.' Suddenly you can feel everyone in the room wanting to water turf. If you're honest and admit you don't irrigate, you get only 3 megs.

'Please examine the survey forms you have to fill out. They need to be as accurate as possible.'

We nod obediently. 'There's going be spot audits,' Garry says sternly. 'Eventually your water use will be measured, perhaps calibrated from your electricity bill.'

Until now, if you'd paid the DLWC the annual fee, amounting to a few hundred dollars, you could pour as much water as you wanted on the hectares you'd been allocated.

'Once you get your allocation we'll move into Stage 2 and discuss how and when you can use the water. The Government has drawn up IEOs – that's short for Interim Environmental Objectives.' Our water use may be going down, but our anxieties are rising.

After Garry has laid down the proposed law, Paul stands up. The spitting image of Phillip's oldest friend, Barry Jones, Paul is talking about Rivercare, an initiative not unlike Landcare that will implement rehabilitation to watercourses.

He announces that it's possible for our association to get a government grant – to hire someone to survey the Pages and provide a 'state of the river' diagnosis on which to base any future repairs. This is the first inspiring piece of news I've heard, but not everyone is as impressed. Someone pipes up with 'If we do a plan, does that mean we have to actually do the work?' Typical, I think. Let's not actually *do* anything. We can get a grant to pay for a survey. We can reveal massive problems. We can agree they should be addressed. And then we can ignore the advice.

Michael, a farmer downstream and a member of the Hunter River Management Committee, puts forward a motion: that a Rivercare plan be considered. I second it. All those in favour say aye. There's a somewhat disgruntled murmur of assent. At least it's a start.

I come home, type up the minutes of the meeting and compose a letter. Everyone needs to know they must fill in their survey forms within a month. Yet most of us hadn't even found the survey amongst all the bumph and brochures we've been sent. The next day I shove a bundle of correspondence into a box for Trish, our reliable postmistress, to drop off on her run.

Our association gets the government grant and soon I'm trying to keep up with Danny, the facilitator, as he strides along the dry

Pages, through Segenhoe Valley, marking his aerial photo for our Rivercare plan.

Here is where white settlement of the Pages began. Thomas Potter McQueen MP, of Bedfordshire, secured a grant of 10 000 acres, complete with six convicts, in 1825. Later he'd add another 10 000 acres to his holding and import twenty skilled men from his English estate, together with their wives and families, under the command of his agent, Peter McIntyre. 'High and forested hills that were the walls of Segenhoe,' said a contemporary account, but they had to hack their way through dense vine brush to claim the valley. (Now only a few acres of 'vine thicket' survive on the foreshore of Lake Glenbawn.)

At the time, Potter McQueen could boast that his was the largest individual investment of any Englishman in the district. But it wasn't enough to save him from being charged with corruption, jailed, bailed out and forced to sell his Bedfordshire estates before fleeing to France, where his wife died soon after.

Nonetheless, the development of Segenhoe continued and, within a few years, McIntyre had spent £18 000 of his boss's money on clearing, fencing, building, buying livestock and operating a flour mill. Segenhoe became a village and McIntyre one of two local magistrates doling out rough justice. Then, like his boss, he was accused of overspending and given the shove.

Deciding to end his exile in France, Potter McQueen leaves his five children with a governess and arrives in Segenhoe in June 1834 to manage his estate and his own affairs. Having borrowed £30 000 from his mother-in-law en route, he is determined to recreate Bedfordshire on the Pages. And his criminal record didn't prevent him from becoming a magistrate and director of the Commercial Banking Company of Sydney.

Five months after arriving in the colony, Governor Burke visits Segenhoe, and Potter McQueen hauls the Union Jack up his personal flagpole, orders a cannon salute and lines up twenty-six men on horses who, augmented with ochred Aborigines, form a guard of honour. Reports of this do-it-yourself ceremony reach London and are reported ironically in the British press. Far from enhancing Potter McQueen's tarnished image, such showing off is seen as evidence of foolishness – which he confirms by having a scandalous love affair. Things go from bad to worse to disastrous. His mill venture fails, debts pile up and after four turbulent years he's forced to leave the district. Sixteen years later Potter McQueen dies a lonely man in Shropshire.

A portrait of him survives, depicting a preposterous man in full military regalia on a white horse. He seems to be emulating Napoleon – or, better still, the Duke of Wellington – whereas, in truth, he was a corrupt, greedy, failed entrepreneur. But he left some fascinating ruins behind him. In 1871, a traveller observes 'stone pillars lying prostrate, bath of marble, and many other signs, indicate what it would have been . . .' Another century would pass before the true agricultural potential of Segenhoe would be realised.

Alan Wood, author of *Dawn in the Valley*, is a direct descendant of a family that shared that Segenhoe adventure. Wood retells stories his mother heard from her mother about the Aboriginal community that lived beside the river. They'd add fermenting wild honey to its water to produce an alcoholic beverage known as 'bool', to be imbibed at ceremonies. There were possums aplenty to hunt and their skins were used to make cloaks and rugs. Strong ropes were made out of kurrajong fibre.

Peter McIntyre described the valley as 'lawless.' In 1826 there were numerous reports of attacks by blacks, and the retaliation of whites was recorded in government despatches. Settlers were requesting more police to stop 'the rule of terror' from spreading, with some settlers recruiting 'friendly natives' to help them keep the 'dangerous blacks' off their properties. But the attempts to co-opt one Aboriginal tribe to fight another apparently failed, as, by 11 September 1826, the Reverend L.E. Threlkeld, who operated a mission in the Lower Hunter, was warning London that 'war had commenced'.

Today, as I stand on the river's edge where Potter McQueen's original Segenhoe evolved, no evidence of Aboriginal life, or death, can be seen. A sandstone cottage, once designated a hospital, is still there, but evidence of McQueen's extravagance has disappeared. We find no broken stone pillars, no marble baths, no hint of gilded halls. On the other hand, we are surrounded by evidence of big, modern investment.

The horse studs of Segenhoe are splendid affairs, offering five-star accommodation to stallions that can cost millions. We pass scores of paddocks, few larger than a couple of hectares, each fenced with the best quality wooden post-and-rail right to the river's edge. Here and there floodwaters have eroded the banks, forcing the fence lines to retreat.

In modern times, the river has become wider and shallower. A few thousand years ago it took a different course, a kilometre away, and the whole valley would accept its inundations, which, in turn, made Segenhoe the fertile place that it is. Of course, no one living along the Pages today is keen on free-flowing inundations; they want more and more of the river banks fenced in, hemmed in. In 2003 a landowner even requested to erect a levee bank. Since we moved to Gundy in 1987 the valley has become entirely formalised with fencing, and the river is expected to behave itself.

Once mullet were abundant and huge perch were so common they were used as pet food. Mussels were found everywhere, even in the upper reaches of brooks. Wild raspberries grew all along the river banks – before sheep chewed them out. Now the river is filled with gravel and silt and much of the water hides underground in a vast aquifer. But it can't hide from the studs' pumps and we pass hundreds of giant irrigators flicking their arched, shimmering, rainbowed water over the lurid green paddocks – water that seems to come from nowhere, conjured up, magical.

We stand beside a fence that's been under water over and over again. The sump oil that's been painted along its length hasn't prevented it from buckling. 'We've moved this back three times,' says the farm manager, looking glumly at a shrinking paddock. And still the river deteriorates, widening as it fills with silt. Sandy beaches one side, coarse pebbles the other. A few metres downstream it's all gravel, about 10 hectares of it, levelled by machines into a plain that now looks like a gibber desert and across which thousands of casuarina seeds have been spread. This is 'flood mitigation work'.

Flood mitigation? A technical term for changing your mind, for discredited theories and failed experiments. This 'flood mitigation' job occupied five men and big machines for two months. The rule of thumb has such efforts costing about $1000 per linear metre. So spread before me is the aftermath of hundreds of thousands of dollars' worth of effort. And I'm told this is perhaps the tenth time the DLWC has tackled it.

More staggering is the sight of poplar tree cuttings, as thick as my arm, being rammed into the gravel. European ferals are being ripped up elsewhere and almost every scientist would say planting natives is best, yet, here in 2003, thousands of poplars have been planted by, yes, the DLWC.

It's a very confused and confusing organisation. Engineers used to be the big wigs, now environmental scientists sit on the high perch. Strategies come, tactics go, changing as fast as wind direction. How can anyone be sure that what they're doing is right? Or even approved?

Further along, monumental blocks of concrete are embedded into the banks, evidence of a fleeting 1950s theory of stabilisation. They were placed, at huge expense, all along the southern end of the valley after the 1955 flood. Once positioned, they were linked with heavy chains in the hope of catching sediment and silt. You can still see a few rusty links protruding from the gravel.

If you drive from Scone to Lake Glenbawn through Segenhoe

you find a spectacular, poplar-lined road. It leads to the Pages River ford that becomes dangerous, then unpassable, as floodwaters rise. On either side of the ford there's evidence of a similar approach to bank stabilisation called 'rock groyne', says Danny, pointing to walls of blocks bound in netting that jut out every 30 metres or so, in the hope of 'catching' silt. And they're fringed by a million weeds, those intrepid colonisers that will, if all goes to plan, be replaced by native trees.

5

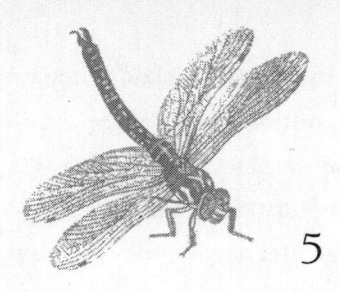

By autumn 2002 our drought is everyone's drought, covering much of eastern Australia and the rest of the continent. It's arrived after three years of wet and seasons described as 'the best on record.' Yet even though we'd expected a dry phase it came upon us so quickly, too quickly. What had been anticipated and feared was, somehow, still a shock.

Everyone is trying to get rid of their cattle. The weaning season is well and truly under way with 3000 head of cattle being yarded each week at Scone. Little wonder prices fall. Others, living on hope, are buying in hay and silage at profiteer's prices, increasing their financial risks. The work and expense of handfeeding will, I suspect, only prolong the agonies of farmers and herds alike. Given past experience, I've sworn never to handfeed cattle again, but that's easy to say. As I fill my backpack to go on a walk at the head of the river, I have to make a decision whether to sell the calves and more cows, or handfeed.

I take a risk. I decide to sell just a handful. The rest we'll keep.

So the pattern begins again where calves, realising they depend on you for survival, lose all fear and would, if you let them, move into the house. Exhausted by sorry attempts at finding feed, cattle simply lie in what little shade they can find. Some will never get up again, but others do at the approach of a truck or tractor carrying hay or silage bales. This sets a bad example for calves that aren't learning about the paddocks as weanlings should. Like indulged teenagers, they're showing no signs of independence.

Correspondence from 1923 describes a desperate local stockman wanting permission from the Pasture Protection Board to 'roast prickly pear on the reserve' by the Pages to feed his animals and to lop a few oak trees *(Casuarina cunninghamii)*. The request was answered in the open, unbureaucratic prose of the era, 'In regard to the roasting of pear, I don't suppose any objection would be taken to that, provided you don't fire the country or spread the pear unnecessarily, but I don't think it would be wise to lop the oaks.' Unfortunately, cutting down trees to feed stock was still practised in the 2002 to 2003 drought.

In conventional agricultural circles it's an article of faith that fertilisers are the best defence against drought and that with regular application of phosphorus, nitrogen and trace elements plants will grow with less water. As everyone in the district knows what others have been doing, we all know which paddocks have been fertilised and which haven't – and can observe the differences now that times are tougher. I am ashamed at feelings of *Schadenfreude* when I see that we're not the only ones suffering and that all the superphosphate in the world doesn't help when it doesn't rain.

Weeks later we send more cattle to the overcrowded abattoirs.

Thankfully, the olive trees are staying with us. We can't sell them. They're really beginning to look like trees and not just bushes. Some are twice as tall as me, yet it's the smaller ones that bear the most fruit. We unplug the drips from the dying and dead and mow over their positions as if they'd never existed, so every row has gaps that, in time, we'll fill. For now I distract myself from the failures by lovingly pruning the survivors.

Forever hopeful, we decide to extend our dryland grove, 6 kilometres away on the other side of the property. More Frantoio and Leccino are planted alongside the Correggiola. So here we'll keep worrying about water and there the trees will suffer benign neglect. But olives are meant to be tough and in Spain, Greece and Italy often seem to do best in the harshest circumstances.

Our friend Andrew, who came here with a portable olive press in 1999, now manages his own nursery and advisory business from Lara in Victoria. During a trip across the Hunter, he arrives with some of his trees and Leandro, a horticulturalist from Argentina who has immigrated to Australia. I show them where the Leccino trees will be planted and Andrew shakes his head in despair. 'You will water them, won't you?' he says. 'Yes,' I say tentatively, aware that his idea of watering and mine aren't the same. Already Phillip and I have hand-watered trees in extremis, but to keep that up would be an endless task, like painting the Harbour Bridge. Andrew pulls out one of his trees from a pot and says, 'This is a strong tree, Patrice. Look at all that root mass. But it needs lots of water!'

I'm back in primary school and feel I'll be sent to the principal's office. This is where biodynamics sees things differently. There's no doubt that a mega-grove, pumped full of computer-controlled water, will grow trees quickly, especially when the water is laced with liquid inorganic fertiliser. But there's something forced and unnatural about them. They're too good to be true. Groves of them remind me of impeccably tidy houses that aren't really homes. Or of artificial flowers in vases.

Despite his reservations about my planting techniques, Andrew offers some extra trees, gifts this time. I'm reluctant to accept them for fear they'll die, or just not grow as spectacularly as Andrew likes. Nonetheless I'm touched by the offer.

Leandro was born in Buenos Aires just 29 years ago. His family moved to Catamarca where he studied agriculture at university prior to working in the olive industry as a researcher. Transplanted to Australia, he's now advising many of the companies planting big groves. As we walk through our tiny grove we talk about the catastrophe that has overwhelmed the Argentinean economy.

'What's the problem with Argentina?' Leandro jokes. 'The Argentineans.' He gives me a tragic smile. 'It's impossible to get ahead there unless you're a politician or a soccer player.'

So he's one of the thousands of Argentineans in a new diaspora, including a brain drain of immense proportions. Leandro talks about the 'thinking class'. 'They're being killed in a different way these days,' he says. 'The intelligentsia isn't being rounded up, jailed or "disappeared", but they remain unappreciated and unable to participate in any rebuilding of the country.'

He speaks sadly about governments with no plans for the future, of not having a strong bureaucracy to keep government on the rails. There are many like him working in the olive industry in Australia, people Leandro knew back in Catamarca where some of the giant, subsidised 300- to 400-hectare groves are being deserted. The jokes I heard when I visited Argentina in 1999 about the rorting of the tax system have come true.

A week later a fax arrives from Andrew giving the water requirements that he and Leandro are advocating. I read it in disbelief and despair. We've been applying one-tenth of the water they're insisting is the bare minimum.

Am I being stingy? Doing something terribly wrong? Looking at our trees I think, considering all the setbacks with frosts and rabbits and cockatoos and, increasingly, kangaroos, they're doing remarkably well. There's something noble about their defiance. Their trunks are thickening; they stand tall and look decidedly happy. They're not being driven into some premature adolescence but are growing at the precise pace olive trees should grow – in tune with nature and tradition. I'm allowing the trees to take their time. Andrew, of course, would argue that I'm denying them their basic needs and forcing them to suffer.

That night departing neighbours give me a 1961 brochure they have found whilst packing, 'This came with the olive trees that were planted in our garden.' 'THE MONEY TREE IS THE OLIVE TREE,' states the front cover. The inside text supports the claim: 'Staggering figures show gigantic increased importation of olives and olive oil.' The brochure promises '20 pounds of fruit from one

Verdale tree'. The nursery, long since gone, was supplying grafted Mission, Verdale and Sevillano, three varieties that are not being promoted today for their oil. The blurb sounds uncannily familiar.

The air is so dry, the grass so dusty, it's hard to breathe. Moisture is a memory. Summer won't give up, instead it seems to be gaining strength. It's throbbingly hot, so numbing and ruthless to vegetation that I'm furious at the wide, blue-white unfriendly sky for crushing all life – mine included.

The olive trees on the hillside grove have green jewels all over them, so (fingers crossed) we'll be harvesting a small crop. Thus far the trees seem happy with the weather, but, after leaving the olives for a few days unattended, I discover the fruit has shrivelled. This is easily rectified with a turn of the tap, by getting the drip system working, but I've been sternly admonished. Don't be careless. Keep your eye on the ball. The first job in farming is to observe, observe, observe.

At least we still have our farm and our olive trees. A friend talks of the plight of olives in the Middle East. Hundreds of thousands of productive olive and fruit trees have been uprooted in the disputed territories by Israel's occupation forces, allegedly to prove that the land is 'uncultivated'. In Gaza alone, by April 2001 the tally was 114 000 trees uprooted, with more in the West Bank, Bethlehem, Jerusalem, Ramallah, Nablus, Tulkarm, Hebron and Jenin.

Here in Australia, we are planting groves in every state, whilst in the Middle East they are being systematically destroyed. International welfare groups seek to fund replanting projects and a letter arrives at Elmswood seeking donations for a grove on Palestinian land: the olive branch, the symbol of peace. And olive oil is expected to salve the wounds of a brutalised region.

Olive trees are burdened with symbolism, and throughout their long history they've been watered by tears as much as rain. The olive sprig in the beak of the Old Testament dove reappears in a sketch by Picasso to become a powerful protest against war. And the olive is also a symbol of endurance, of defiantly clinging to life. You even hear the argument that Australia, with all its droughts, is too benign a place to grow the best olives – which survive, even thrive, in far harsher conditions. Our trees are unlikely to be uprooted by armies, or deserted by global corporate pragmatists. Ours have to deal with El Niño, the kangaroo, the cockatoo and grower fatigue.

In 2001 an excess of rain had destroyed much of our olive crop. We learn from that experience to better judge the optimum time for harvesting, so we check the fruit every day to detect any hint of changing colour. I want to hold out until the fruit in all sections turns black and ripe – given increasing evidence from other harvests that picking too early makes processing difficult and less productive.

I'm not alone in the guessing game. Many groves are harvesting sections each week, sending the oil for scientific assessment of acid and peroxide levels, to ensure optimum oil. Labs have backlogs of samples. And every oil tested or tasted is subtly different. It isn't just a question of timing; regional differences are also emerging. The east coast, with its greater humidity, is producing oils that are different in taste and colour from the products of South Australia, Western Australia and Victoria.

We still have some oil that was processed at Elmswood in March 1999 from the wild olives we'd picked from ancient, abandoned trees. As I decant some from the stainless steel vat in the cellar it remains powerfully aromatic, but the flavour quickly fades. (It looks fresh, but green doesn't guarantee it's good.) Nonetheless, even the last drop is better than the oils available on supermarket shelves, largely the bottom of the European barrels foisted on unsuspecting foreigners. I learn that storing oil requires airtight containers and cool, dark places. Even then, buying bulk oil is a false economy.

Olive oil, like bread, is best fresh.

6

Sitting at the kitchen table, sipping a mug of tea, I browse through the PR handouts masquerading as a local newspaper and come upon a story about a couple recycling an old Cobb & Co. coach-house at Wingen. It seems that no sooner had they moved in and started operating a B&B with impeccable heritage connections than they'd made a surprising discovery. Their ancestors were buried nearby, on the edges of the Pages River, north of Cameron's Gorge.

Their dynastic name is well known in both the pastoral and literary worlds: Wright. This generation of Wrights, Rick and Nora, left their landholding in the New England region after a monumental financial battle with the banks. Rick's cousin was the late poet, Judith McKinney Wright. In her book, *The Generations of Men*, Judith tells the epic tale of the life and times of her grandmother, the clan's matriarch, and describes the triumphs and despairs of pastoral life since the 1840s, including the era when the Wrights took up their first few thousand acres at Bickham, along the Pages River south of Blandford.

I'm sorry I never knew Judith Wright. Phillip and Judith corresponded for years – they'd been introduced by Nugget

Coombs – and were still exchanging enraged letters about John Howard a few days before her death. But for me she remained a mythological figure and the first famous person (and the only poet) who'd come to St Aloysius College in Adelaide during my last year.

She'd read her poems to a room full of teenagers in the hope of inspiring us to engage in the worlds of literature and, yes, politics. But I'd been unimpressed. Like most of the other students, I'd seen just an old-fashioned lady with a strange, grating voice, and hadn't made much of the poetry. Yes, this is an appalling thing to admit, but that's a seventeen-year-old for you, judging people by appearances. Judging a poet by her cover.

Recently I decided to make amends by reading all her work, only to find that her books were out of print and none were on the shelves of our local library, nor could I find them in Phillip's, as he'd recently done a massive reshuffle and hadn't memorised the new layout. So I had to order every one and ended up tracking down a first edition of *The Generations of Men*, paying less for it than a newly released novel. As I turned the pages, I discovered a kindred spirit in her arguments about 'values' versus 'science'. Arguments I could have used to boost mine when having rows with Phillip, who insists science has the answer for everything.

On a breathlessly hot day, Phillip and I meet up with Michaela, the present owner of Bickham, to visit the Wright family graves. Hidden below the hill and guarded by a rock escarpment called the 'Sphinx', a kilometre from the Bickham homestead, was the last resting place of five Wrights. Buried between 1889 and 1954, they lie surrounded by a circular dry stone wall 10 metres across and 2 metres high, with a heavy, hand-forged metal gate. It's a short distance from the river, on the edge of a paddock that, when examined on the map, looks eerily coffin-shaped, adding to the sense of it being ancient, a small necropolis, abandoned by a forgotten civilisation. Also in the coffin paddock, as Egyptian as the 'Sphinx', are massive pyramid-shaped rocks.

The basalt rubble construction of the Wrights' last resting place was designed by the prolific architect John Horbury Hunt (1838–1904). Hunt was responsible for many, if not most, of the architecturally impressive buildings along the Pages Valley, including St Paul's Anglican Church in Murrurundi, St Luke's Anglican Church in Blandford and Glenalvon Stables in Murrurundi.

After 113 years, the enclosure is almost smothered in ivy, hiding the castellated coping, though cattle, munching at the leaves when we arrive, had pruned it as high as they could reach, revealing both the scale and quality of the construction. Alongside the gate are two huge rough-barked apple trees (*Angophora floribunda*), which must have been strategically planted many lifetimes ago. Now they are dismantling the place, twisting, turning and heaving their branches over the gate and graves, where moss and lichen obscure the names and dates chiselled into the stone slabs. Nothing except the fact that the Wrights are buried here is revealed in this impressive, secretive and mysterious site.

Is that the growl of an approaching storm? The rumble of an earthquake? The sound of bombs falling on a distant battlefield? It's all of these things and more. It's the ominous sound of that dreadful rumour that's gained in strength until, now, suddenly, it's 7.5 on the Richter scale and the earth beneath me is shaking, cracking, collapsing.

I learn that bulldozers are gouging roads over the hills and holes are being drilled all over the place for core samples; and a phone call to the New South Wales Department of Mineral Resources (DMR) at Singleton confirms that, yes, an 'exploratory licence' has been granted – covering an area of thousands of hectares surrounding Blandford, with the Pages River flowing right through the middle of it.

Officials make soothing noises. It's only early stages. By no means certain. Might come to nothing. Not to worry.

There's nothing more alarming than such soothing statements. Phillip and I look at each other, not in disbelief, but belief itself. A phone call in February 2002 to David Foster, owner of the company conducting the tests, confirms it.

'Have you found coal?'

'Yes. But the seam is on an angle.'

'Will you be mining it?'

'I hope so.'

'Will the mine straddle the river?'

'We don't know yet.'

'But we've spent months trying to work out a water-sharing plan.'

Silence.

'Don't you think you should come to our meetings?'

'Patrice, we don't expect to be taking water from the river. We might even be putting some water into it.'

Stay calm, I tell myself. This is, after all, the Hunter Valley and there's drilling going on everywhere. Over the years a number of fossickers or geologists have knocked on Elmswood's door, telling us they've a licence to look for minerals on our property. They'll roam the hills for a few days, tap at a few rocks with their little hammers or dig a few holes. Then off they'll go and everything's forgotten.

However, it's impossible to remain in denial when a neighbour phones to tell me truck drivers at a long-established mine down the valley are being recruited (contracted) for a new mine at Blandford. Within days I hear that David Foster's company is purchasing farmland for, yes, Bickham Coal Company Pty Ltd.

And not just any old land. They have purchased a property called South Bickham, a 3000-acre portion that, until 1982, was part of the property the Wrights had established in the 1840s. How did Michaela feel about this? A mine right next to her house – the boundary literally a hundred metres from her back door. A mine with the same name as her farm.

So there it was. A major ecological issue in our backyard. This was not some distant issue, like uranium mining in the Northern Territory. This was personal.

I went to bed telling myself not to oppose the coalmine. Well, not yet. After all, I use more than my share of coal-generated electricity. We haven't converted Elmswood to solar or wind generation. When I turn on my computer, I use electricity from the burning of coal that's been ripped from the Hunter Valley. The metallic components in that computer may have been made with the assistance of coking coal exported from the Hunter. When Aurora plays a CD or I turn on an air-conditioner or electric pump to irrigate a crop, we're using electricity generated in the power stations we can see smoking and steaming from the top of Black Mountain.

Here at home, giant electricity pylons march over the back of Elmswood, stitching the hills with their cables, linking Tamworth to Liddell Power Station, south of Muswellbrook. Erected in the 1960s by Transfield, which employed teams of Italians for months, the pylons are our procession of steel robots carrying that constant pulsing current to support our modern lifestyle. All along Segenhoe Valley they scallop Brushy Hill, crossing the Pages, before making their way over Black Mountain and eventually to the Pages River below the Bickham Mine site. In a hundred years or so, they'll be rusting relics, to be admired like the ruins of Roman aqueducts.

Furthermore, I'm not a NIMBY. I've always had contempt for the 'Not-In-My-Backyard' approach to environmental issues. If not in mine, then whose?

Like other farmers in the district, we curse the mines for stealing our workforce and inflating our wage bill. But it has to be said that, again and again, the mines have saved many a struggling farming family by giving father or son that extra off-farm income. Instead of working eighteen-hour days for next to nothing, you can get an average of $80 000 per annum for driving a truck or overseeing machinery.

Every tier of government wants coalmines: federal, state and local. Which is why Big Coal dominates the valley. Not since the heyday of the timber industry has there been such an economic boom. Once upon a time the industry was centred on Cessnock, but now it is Singleton and Muswellbrook that are the principal beneficiaries – having the highest per capita incomes in rural New South Wales.

Yes, there are a few small problems with coal – like the greenhouse effect and global warming. Like the global crisis the Kyoto Protocols are trying to tackle. It may be fossil fuel that adds more greenhouse gases than any other energy source, but the world is addicted to it, and Australia – the Hunter – is one of the major 'pushers'. No wonder that Australia, like the USA, snubs Kyoto.

There are still some quisling scientists who run the coal-industry line: that greenhouse is just a theory and that global warming, if it's occurring at all, is probably a natural cycle. But they sound like Creationists contesting the theory of evolution. Global warming is happening and the overwhelming consensus is that much of the trouble begins right here in our Hunter Valley.

A sleepless night confirms that I can't accept slippery economic arguments. A coalmine on our river? We need another great hole in the ground like a hole in the head.

A coalmine is a corporate entity that believes that its rights are greater than those of entire communities. As corporations go, coalmining companies are particularly cocky. And why not? In the 1980s it was discovered that, through a legislative anomaly, some landholders actually owned the coal rights on their properties. So, in 1981, the Wran Government enacted the *Coal Acquisition Act* to ensure that *all* coal became 'the people's coal' and the government ensured that each tonne of coal would provide Treasury with revenue – countless millions of dollars per annum.

So what is under the ground is deemed far more valuable than what's on top, whether we're talking beautiful landscape, biodiversity or agriculture. Entire valleys of dairy farms have disappeared as the giant draglines have advanced. Even the Hunter's most prized and prestigious industry, wine production, has been pushed aside – as was seen recently in the famous case of Rosemount Wines versus Bengalla.

Before being swallowed up by a public company, Southcorp Pty Ltd, Rosemount was a family business, owned by the Oatleys. One of the best-known vineyards and most respected wine labels, Rosemount took the Bengalla mine to court, claiming coalmining would jeopardise, perhaps destroy, their business and that, in the long run, bunches of grapes would be as important to the economy as lumps of coal. The court agreed. But no sooner had the Oatleys toasted their victory in local bubbly than the New South Wales Labor Government introduced overriding legislation – ensuring that greenhouse would triumph over chardonnay. I knew this would almost certainly happen again – that even the smallest of victories over Bickham Coal Company Pty Ltd would most likely be pyrrhic.

'Negative impacts' can't undermine mines. They become 'acceptable risks.' The New South Wales Department of Mineral Resources, the bureaucracy responsible for overseeing – more accurately, encouraging – mines, was forever lowering the bar.

I ask landowners north of us for permission to cross their property to enter the southern end of Cameron's Gorge on the Pages. Somewhat reluctantly, they agree. With picnic supplies in our backpack, Phillip, Aurora and I take off for what we think will be a leisurely afternoon walk. But on this hot drowsy day it is a taxing clamber over rocks that takes us only a small way into the gorge, and progress is slowed by Aurora's determination to strip off and swim in every rock pool. In one she finds herself sharing the water with a red-bellied black snake.

Back in 1972, the National Parks and Wildlife Service suggested that Cameron's Gorge, 10 kilometres upstream from us at Elmswood, be gazetted as a nature reserve. Described as an area 'surrounded by a sea of cleared agricultural land on the headwaters of the Hunter River,' scientists saw it as being important to conservation and, by 1975, an acquisition strategy for the purchase of surrounding land was in place.

The ecosystem falls within the splendidly named Cranky Corner Landsystem and no other cranky corner land in the National Parks portfolio has such a riverine environment. In November 1987, 1280 hectares was officially dedicated as Nature

Reserve and placed under the management of a much-maligned New South Wales National Parks and Wildlife Service. During the time it was being assessed, bureaucrats in another government department had a completely different plan for the area. They wanted to dam it.

Glenbawn Dam, now Lake Glenbawn, on the Hunter River, was already storing water and there were plans afoot to increase its holding capacity. But there were also hopes of finding other places to store water, and the Pages provided possibilities. One plan was for a dam wall to cross the Pages just before it enters Segenhoe Valley, at the northwest tip of Brushy Hill, thereby flooding our Gundy Road and most of the township of Gundy. It would have brought water to Elmswood's front door, if not over it. The view from our veranda would have been spectacular.

According to the latest mapping of native vegetation, Brushy Hill, a long escarpment of limestone, is one of the most diverse and important areas in the Hunter. From a distance, the area looks unremarkable; it wasn't until I was taken deep into it that the profound differences were revealed. Vine thickets and unexpected plants like spinifex cling to the hills. It frames Segenhoe Valley to the east and Lake Glenbawn to the west.

Near the proposed dam wall, a lime mine operated throughout the 1940s until 1951. Driving along Gundy Road you can still glimpse a rusty chimney, marking the ruins of the enterprise. And if you look hard beyond the regrowth, it's possible to see where the limestone was dynamited. Old Department of Mines documents record that seven men used a 'jaw crusher, hammer mill, and gas engine' to convert the rock into agricultural lime. Allegedly the operation closed when they ran out of suitable timber for the production of charcoal that was needed to fire the gas power plant. Moreover, the flooding Pages River was a problem.

The mine was small and most of its lime was trucked to fertilise the mine-owner's property on the outskirts of Scone. Locals remember his whole hillside caked white. The remains of the plant, still on leased Crown Land, are a sad, sorry cluster of bent steel, shrouded by trees and full of the stench of dead animals. For years it's been used as a graveyard, and the bleached bones of cattle, horses and sheep are piled high and choked with weeds and ancient wool.

Just before the river enters Segenhoe Valley, below the lime mine, where the river begins its secret, underground journey into the aquifer, is a Travelling Stock Reserve known since 1913 as 'Bobbies' in honour of 'Yellow Bob.' After retiring as a stockman, Yellow Bob lived alone here in a humpy. His birth name is Robert Stephens and, although photographs clearly depict an Aborigine, it is believed he was half-caste. According to folklore, Yellow Bob found his retirement years demoralising. I suspect that his presence along the river became an embarrassment to the local whites, who had police remove him from the reserve. He ended his life in an 'institution.' I've been unable to find out if this was a jail, an asylum or a nursing home. Records, reeking of paternalism, talk of him being given decent clothing and nicer surroundings. Yellow Bob's mystery is characteristic of how local Aboriginal history is fudged, forged and forgotten. While we preserve our stories through genealogy and photograph albums, the stories of the Aborigines who lived along the Pages are not only gone but deliberately suppressed. The three surviving photos of Robert Stephens record the only known indigenous person from the past who called the Pages home.

The road from Scone now cuts through rock along the Pages on the other side of the river, replacing four river crossings that once provided access. On the far side of Bobbies Reserve, a wine shop owned by the Moran family, whose grandson Trevor did Elmswood's pruning, once did good business supplying the drivers of bullock teams and other travellers. Though the community of Brushy Hill is long gone, remains of homes and a school still survive.

A travelling stock route still crosses the river; you see old white wooden signs attached to gates pointing drovers to waterholes for stock, marking the track as it winds around the cuttings and river bends. The river remains the main route for drovers and stock seeking feed in difficult times. This is the way it has been ever since white settlement, yet our new water-sharing plans ignore this tradition – suggesting that eventually no stock will have river access.

With plans for Brushy Hill Dam quashed, another idea was to build a wall across the Pages at the northern end of Cameron's Gorge. This would have created a 200- to 250-gigalitre dam, providing the people of Blandford with a wide river and lake-like feature, flooding numerous productive river flats along the way.

In the end, Glenbawn's wall was extended to treble its capacity and the 'little' dam ideas remained damned, amongst floods of reports, in the DLWC's filing cabinet. Since Glenbawn opened in 1958, its prime purpose being to help 'industry in the valley', it has become one of the shire's main tourist attractions, with 80 per cent of its revenue raised from recreational fishing. After years of continuous releases by New South Wales Fisheries, Lake Glenbawn promises a good chance of catching a 'mega bass'. In tourist brochures and on posters, this modest fish looks as big as a Murray cod, if not a whale. Whereas before most traffic on our local roads seemed to be floats and trucks servicing the horse industries, these days boats and fishermen heading off on piscatorial encounters are just as common. At our local pub you're bound to overhear discussions of what kind of bait, tackle and boats are best for bass catching.

In March 2001, after months of preparation, our tiny township has its 'Back to Gundy' weekend. A large white sheet becomes a makeshift screen in the Gundy Memorial Hall for a screening of *The Shiralee*, starring Peter Finch. This movie, shot around the township in the mid 1950s, kicks off the two-day celebration that has old-timers returning to tell stories, leaf through photo albums and remember floods and fires, wars and depressions.

We invite Helen Brayshaw and her family to visit us for drinks before the opening night. She knows Elmswood well. Her grandfather once owned it and she grew up on a neighbouring property, Miranee.

Not long after we purchased Elmswood, Phillip discovered Helen's *Aborigines of the Hunter Valley* in the Scone newsagents. It contains 19th-century sketches of elaborately carved trees where the Pages and Isis converge, soaring evidence of Aboriginal occupation. But an excited trip to the site shows no sign of them. Helen's uncle remembers the last two trees and how he'd climbed them as a kid until a flood finally bowled them over. They were the last survivors of 120 carved trees – a forest gallery – that stood there long before Elmswood.

'This area must have been a very important place for Aboriginal people,' says Helen. 'People probably came from many parts of the Hunter Valley to meet old friends, exchange stories, just like us now at the "Back to Gundy" weekend.'

Working as a heritage consultant, Helen arrives with a facsimile of an 1828 map, one of the oldest of the district, usually referred to as Stewart's Map, after John Stewart who owned the original grant for Elmswood. The map centres on the convergence of the Isis and the Pages, where our house has stood for over a hundred years, but on the map the Pages is called the McIntyre.

While we study it, Helen mentions that she conducted the archaeological survey of Cameron's Gorge for the dam proposal and established the existence of at least two significant Aboriginal sites at the north end of the reserve – and many others that would have been submerged had the dam gone ahead. I suggest we go for a hike to relocate them. She readily agrees.

So weeks later, at 9.30 on a crisp, clear winter's morning, Helen and I set off from the southern end of the gorge with cameras, maps, GPS (Global Positioning System, a hand-held personal navigator about the size of a small camera) and trudge north. There is no 4WD track and the scrub is too dense for horse riding. No sooner have we left the truck than my ankles begin aching (clearly all my walking hasn't made me any fitter), I nearly fall into the river and doubt I can keep going. But after a couple of hours, much of it spent climbing over massive boulders we can't squeeze around, it becomes a case of push on regardless.

Fit and organised, Helen marks the map to gauge the distance we're making and, having just returned from a trip to Central Australia, is now an expert with the GPS. Though I long to linger, she keeps pushing me on, around or over rocks often more than 20 metres high and looking like they have been art-directed for a Tom Roberts painting, except that cormorants are perched on many of them, painting the rocks white with poop.

We see no fish in the deep, turquoise pools. At least the pools are free of the curse of carp that have invaded the Pages at Elmswood. I'm led to believe that everything around us is safeguarded – forever a sanctuary for all the unknown critters to live uninterrupted. And all of this is just upstream from us. Here's a perfect filter for the river, an opportunity to give it a second chance before it approaches the aquifer at Segenhoe. The reserve, as one National Parks ranger put it, 'holds the status quo'. The gorge is public property, its main purpose the 'preservation' of nature for scientific and educational purposes. However a 'permissive occupancy' lease still applies, which allows cattle to graze – so much for being a Nature Reserve.

Over the past few years the National Parks and Wildlife Service has bought numerous blocks of land in the Scone Shire, designating them as parks or reserves, though not all are quarantined from humans. Not far from Cameron's Gorge is the Burning Mountain Nature Reserve, where a coal seam has been smouldering for thousands of years. Although it occupies only 15 hectares, it has thousands of visitors a year sniffing the sulphurous stench. The pathway leading up to the burning seam offers extensive views of the Pages to the east, where it snakes around hills on its way to the gorge.

When we can, Helen and I leave the hard rocks for the couch grass, now so common that many assume it's a native. We startle a few cattle and an occasional kangaroo, wallaroo or wallaby. Pools of water look certain to house platypus, but we don't see any. Under a cloudy sky trying to dump rain, we break for lunch and take a GPS reading. Helen presses some more buttons and a GPS screen draws a little map showing our truck is 4.73 kilometres away as the crow flies. It has taken us three and a half hours to get this far.

We're in a drought and I want it to rain, but not today – not while we're clambering over the giant egg-shaped rocks; not while we're eating our boiled-egg lunch and Helen is leafing through her original report with its colour photos, so that we might more easily identify the Aboriginal sites. But the trees have grown, some areas are no longer cleared and it's not so easy. Walking up and around a gully, we recross the river and discover Aboriginal stone artefacts. This is the first time I've ever seen them in situ. And as we walk, eyes to the ground, up the river, next to a fence, we come across more and more and more. Just as they were twenty years ago. Just as they were perhaps 20 000 years ago.

Why here? Why aren't we finding these artefacts closer to the river? 'We would if we looked,' says Helen. 'But the river dumps silt and we'd have to sift through a lot of dirt.' All along the Pages, if you take the time, you'll find proof of Aboriginal occupation. (And months later as I walk with my retrained eyes I find them all over the place.)

To demonstrate, Helen reaches down and picks up another artefact, places it on a rock, puts her lens cap next to it to give it scale, and takes a photo. Had we been permitted to begin here, we'd have had all day to look. Now we need to leave soon if we're to make it back to the truck before dark. And before rain.

It's good to know that the archaeological record remains undisturbed, that it hasn't drowned under a dam, been wrecked by development, buried, removed or destroyed. Yet protection of cultural artefacts wasn't the reason the gorge was made into a reserve. Its Aboriginal history wasn't a factor in the decision.

Now Cameron's Gorge is one of the Pages River's secrets. While it's ludicrous to compare our little stream with the Nile, for example, I couldn't help but think of connections. For if you trace the Nile far, far upstream you'll come across a gorge in Ethiopia that remained unknown to Western explorers until the 1960s, when a US survey team descended into it and found ancient mummies stashed in its caves. Alan Moorehead tagged along and retold the story in his great book, *The Blue Nile*.

There's comfort and pleasure in the knowledge that the Nile, despite having been described by Herodotus in 640 BC, still had its secrets. And along the Pages in the 1980s the Aboriginal artefacts Helen found, dating back possibly 10 000 to 20 000 years, are vastly older than the Blue Nile mummies, and comparably significant.

Helen wears sturdy boots that allow her to wade through the low flow with ease, whilst I take off my sneakers to cross, happy to cool my feet. For a while I walk around the grass barefoot and pick up ancient artefacts, placing them back precisely where I found them.

We increase our pace, retracing our route, crossing the same sandbars, passing the same native fig espaliering across a vertical cliff, slowing down when we feel a change of mood, often when signs of white occupation become more evident: a peppercorn tree, an old fence or the remains of a holding yard. As we gaze down into pools of delicious depth and clarity, I rejoice in the fact that no dam was built and that at least one of the pages of the Pages hasn't been smudged or torn.

The following morning Helen and I wish we'd rented wheelchairs. My legs are so stiff I have to grip the rail on our staircase to get downstairs.

There's one part of the river I try not to think about: a neighbour's tip. Tonnes of junk and rubbish have been dumped into a crater beside the river. It must have been there for a hundred years, filling up with the flotsam and jetsam of farm life, but recent additions have more than doubled its size. It's piled high with wire netting, old fridges, freezers, tyres and plastic chemical containers, positioned so as to be invisible to its owners. Unfortunately I see it every day when driving from one side of Elmswood to the other. I try to edit it out of my mind; nonetheless it destroys the beauty of a long stretch of the Pages.

Tips are the ugliest of archaeological digs. When we arrived at Elmswood, we found two in dry gullies on our property and spent a fortune filling a convoy of trucks that headed into the official municipal tip in Scone. To have one facing onto the Pages just adds to the disgrace. Over the years we've complained to the neighbours about it. Often. Nothing happened. We've spoken to the Environment Protection Authority (the EPA) and, once again, nothing. We've written to the DLWC. Nothing. Everyone agreed it was appalling, but everyone declines to act.

In the past we could see only the tip of the iceberg (the tip's tip, so to speak). Now, as the shrubs and grass around it die in the drought, it emerges ugly as a festering wound and, as the final insult, Phillip discovers corpses of recently dead horses, which feral pigs were tearing at, just feet from the water's edge and a few hundred metres upstream from where the residents of Gundy draw water. Outraged, Phillip goes home, grabs the camera and returns to take photographs. Yes, horses die. We too have livestock and, when our cattle die, it's sometimes appropriate to move them. But to dump them by the edge of a river was, to put it mildly, provocative.

'Don't write one of "those" letters,' I beg Phillip. Words are his favourite way of waging war and whether the target's a politician, a corporation or a neighbour, the collateral damage can be significant. 'They'll never talk to us again,' I shout. Exhausted by the drought, I'm willing to put up with the tip, having done so for years.

But not Phillip. He grabs his dictaphone as if it were a six-shooter and starts firing verbal bullets. Weeks later the tip is cleaned out and now you'd never know it had been there.

7

Magnifying glass in hand, I bend over maps spread across the kitchen table. I'm counting gullies, trying to find exactly where I was struggling up steep hills earlier today. It's the sort of activity my mother would have considered a total waste of time, like my determination to know plants by their Latin names.

In her final years, Mum had actually complained about my commentaries when we walked around the garden, pleading with me to simply be quiet and enjoy the experience rather than make it quasi-scientific. For much of her life she had to endure my fascination with maps. I took them with me everywhere. For me, maps validate a journey; for her, they were just scratches on paper. Moreover, maps are to farmers what gardening books are to gardeners, indispensable in planning and thinking. I understand how William Smith, the hero of Simon Winchester's *The Map That Changed the World*, could bend over his table for twenty years, working on a single cartographic masterpiece.

Now maps crowd both my office and the kitchen, rolled up in cardboard tubes or uncoiling on chairs. But Phillip can't complain because the house is so cluttered with his collections of everything. It'd be different if I never looked at my maps, then they could be exiled to the attic. But every other day there's a reason to pore over one, usually on the kitchen table – like I'm doing now. I squint at a thin blue line representing a particular gully and know the physical truth of what it signifies.

One day, a man who knows much more about thin blue lines on maps than I'll ever know knocks on the kitchen door. His nut-brown face is fringed by a silver beard, making him look like Abraham Lincoln. He asks permission to walk all over the farm, to map it in the finest detail. More than that, he wants permission for hundreds of others to walk all over the farm as well. He's organising the 2000 Australasian Rogaine, a long-distance cross-country navigation event that will attract entrants from all over the nation and the region.

Hundreds of strangers walking over our farm? The place we bought primarily for privacy? I don't think so.

However, the arrival of Graham, rogaine ambassador, is a gift. Such is his natural charm, a quality of innocence, that you can't say no – although, at the beginning, I tried to. For Phillip, it was love at first sight. He thought Graham was fascinating, the project harmless and so I was quickly talked around. During the following eighteen months, Graham traverses the river and our hills, plotting a course that will make the rogaine both challenging and safe.

The difference between a rogaine and a good bushwalk? Rogaining is a sport, in the truest sense. Participants work in small teams and the objective is to navigate by map and compass and, at night, by the stars, to as many carefully hidden checkpoints as possible. Each checkpoint is valued according to the difficulty of finding it, and the trophies go to those who plan their journey most efficiently, reaching the most highly valued checkpoints in the least time.

It's a race requiring a high IQ as well as physical fitness, so many rogainers come from universities, like Graham, who works at the University of New England's Physics Department fixing scientific gear. But his after-work hours are devoted to walking and 'thumbing' maps, as he called it. Thumbing is the rogaine term for holding a folded map with your thumb on your precise location, so you can look at the map and say 'That's where I am'. So as you walk, your thumbnail crawls along the blue lines.

On one visit, Graham was carrying a GPS gadget. It was the first time I'd seen one, let alone held one. Mum wouldn't have approved. She'd have seen it as another unnecessary gizmo. (I think her thoughts live on in Aurora who, as a six-year-old, declared that she could never get lost because she'd 'always be with myself.' For Aurora, as for Mum, we are simply where we are.) For Graham, the GPS meant he could be absolutely certain that his maps were accurate.

One sweltering Sunday afternoon, Aurora and I, having given up on cooking, are lounging around reading when a twenty-something bloke knocks at the door. Introducing himself as Greg, he wants to know if we own the block of land down the road on the other side of the river.

'Why do you want to know?' I ask suspiciously.

'Well, that's where 32 degrees South 151 degrees East is. It's a latitude–longitude integer-degree intersection. I'd be the first person there.'

I didn't like to spoil his fun by telling him a few of us had already beaten him to it.

He writes out a web-site address and suggests we look it up. 'Is it your place?' he asks again.

Again I hesitate. 'What will you do when you get there?'

'I'll stand there and take the GPS reading, then a digital photo, and put it straight on the web.'

Having given permission, Aurora and I go back to reading and then later check out www.confluence.org and discover that, yes, there are people with GPS devices in their pockets tramping all over the world – locating the integer-degree intersections. New points are always being reached. There on the web site is our seventh paddock, and the views from it, facing north, facing south, up close, long distance. Greg had made it to 32 degrees South 151 degrees East, 272 metres above sea level. He was there – on our place, showing the world. So it's not just William Smith, rogainers and me, but a whole population of mad mappers and confluence hunters out there animating thin blue lines.

So www.confluence.org creates a network of people joining together for a big idea. It's no longer enough to simply explore and record your experience in a journal that might one day be published. It's all now, now, now. Thanks to cyberspace, these particular places in time and space can be instantly proclaimed. I can see 'confluence reunions' in years to come, when the participants, grey and aching, remember the good old days and finally meet face to face, rather than in cyberspace.

Months after sending off our water-licence surveys, we're all given our water allocations. The DLWC has granted a total of 4610 megalitres a year, to be divided among eighty-five licence holders, and allows a further 5000 megalitres to be taken from the Segenhoe Aquifer. That's 9610 megalitres from the Pages, plus 1653 megalitres from the Isis, the Pages' main tributary. As well, there's an unlimited and uncalculated extraction from hundreds of people who are entitled to take their 'basic water right' for domestic use. It sounds huge. And it is. Agriculture uses vast amounts of water – 72 per cent of all the water taken from all the rivers. And farmers, always being told to modernise and seek new markets, want growth and diversity, but that's impossible without water.

The only people facing the possibility of real restrictions on the Pages seem to be those who wish to continue with irrigation. The others, who use river water for stock and domestic use, have priority and take what they want, so long as the water's there.

Some irrigators are referred to as 'sleepers': they have an allocation, but don't use it. Many are retired, others just tired. If you use only a small part of your entitlement, you're a 'dozer'. It's clear that the river has never experienced the full take-up of its allocations, but now, with a moratorium on licences, the sleeper and dozer entitlements are waking up.

The new water-trading arrangements may actually lead to more water being used, not less. And here's where the DLWC has to cop a huge amount of criticism. Why on earth did they ever allocate so many licences in the first place? Why didn't they, as keeper of the watercourses, show a glimmer of environmental concern years ago?

It becomes obvious that all the changes being proposed are based on hunches, bets, theories, hypotheses, guesses, hopes. And the one interested party, for whom all this is being done, the river, can't get a word in edgeways.

Two months later we meet again in the Gundy Memorial Hall and learn that some people have fared well and others badly. Why weren't all the licences converted on the same basis? In other words, regardless of what the water is used for, simply multiply a hectare by 6 megalitres. But that didn't happen. Consequently, we have a hall full of disgruntled landowners – some fourth-generation farmers – facing a dramatically reduced entitlement. The process seems gratuitously complex. Letters are drafted to send to an anomalies commission for consideration.

We now move on to Phase 2, to the big issue of 'when' we can take the water. Nothing matters if the river is flowing strongly. The issue has always been, and always will be, who gets water when it's low. Unlike the regulated rivers where dams provide the give and take, on an 'unregulated' stream we rely on nature.

The DLWC argues, quite logically, that it's better to harvest water during times of high flow. But there's a catch. It has to be pumped into purpose-built dams, not into dams that already catch run-off.

So Phillip and I decide on two things. Yes, we'll build a dam and fill it when the river is running high, so we can leave the water in the river when it's low; in that way we'll be able to keep the olive grove alive. Secondly, I'll attend the next WaterWise course, one of the best hands-on initiatives to help farmers measure their water use and become more efficient.

But first the dam. Deciding to dig one is easy, but where to put it is another matter. All along the river, people are talking about the necessity to build appropriate dams – only to discover they've no place for one. There are already regulations preventing farmers from building too many new dams that cut off water to the river. If your farm is small, it may mean you're out of luck – that the DLWC just won't let you dig one. A legal one, that is. But the issue for Elmswood is that we need the dam to be close to the existing infrastructure. What is the point of a dam if the water can't be used for our olives?

It has to be beside and below the 30 000-gallon tank that gravity-feeds pipes to the grove. Even that means special diesel pumps, with filters, to lift the water uphill again. Then there is the question of how to harvest the river during high flows. Would we need a whole new pumping system? Or could we manage with the small, existing pump?

If we were starting out and designing a water system from scratch, we'd do it differently. But what we have in place is tens of thousands of dollars' worth of underground pipes leading to four tanks on three widely separated hills. So choosing the place for a dam becomes as critical an exercise as working out where to plant the olives – and that had taken months. Finally, we settle on a gully between the tank and a dry creek. I phone Steve at the DLWC and ask him to assess the site. Days later he arrives with the perpetually smiling Col, who will oversee the work, and together they spend a couple of hours lugging laser-surveying equipment and agree a dam holding 13 megalitres would fit quite neatly across the gully.

'Can we take that tree out?' Col asks.

'No, make the dam curve around it.'

Leaving pegs in the ground to mark the edges of the proposed dam, they take away samples of soil and water from the river. If the water is too saline, it will react to the soil, making the clay more porous and therefore unlikely to seal. They warn me that the whole process will be pretty expensive, but both Phillip and I warm to the thought of a big dam – and not just for drought protection. It will, in time, look marvellous. It will attract birds and turtles and give coolness to a dry gully. Already Aurora envisages herself canoeing around it. When the river is dry, there'll still be some water to play with.

Col's been busy measuring proposed dams across the district. Everyone has the same idea – get a dam in now and be done with it. This sort of water storage is going to be the new way of doing things. Then, while we're waiting for the tests and the arrival of the giant D5 bulldozer, another idea comes along completely out of the blue: a private irrigation scheme, a giant pipe from Lake

Glenbawn passing right past our place, giving the village of Gundy and farms like ours unprecedented drought protection. It is, as yet, pie in the sky, a long-term proposal – and there's no guarantee it will come off.

Ancient, gnarled *Angophora floribundas* fringe the river bank around our mono-pump. In their fight for breathing space, they tear each other apart and limbs fall to the ground, exposing rot deep within their trunks. One tree literally splits itself in two, its woody guts revealing terminal disease. It wasn't good enough to simply collapse. It had to take the telephone cable with it. Now, dangling just a metre above the river, is the phone line to Moonan Flat and beyond.

The slightest flush of water down the dry bed will snap the cable and disconnect a lot of neighbours, so I phone Telstra. Or rather, I try to. Like the phone line across the river, I'm left dangling, listening to canned music and recorded voices that infuriate me. Eventually I tell someone, somewhere – perhaps in a call-centre in Mumbai – what has happened, stress the urgency of the job, and leave my phone number.

No one calls back and three months pass. I go through the whole rigmarole again and leave the same message. Once again, the silence from Telstra is deafening. Neighbours reassure me that the cable must be obsolete, so I forget it. Finally, four months after my original call, five men arrive to repair the cable. Had my calls reached the right department? 'Dunno,' says one of the blokes. 'We've only just heard about the job.'

Years ago the same cable had dangled six metres above the river – and we'd watched as the floodwaters rose to destroy it. Up and up the waters came, dark with detritus, until the cable began to flail, to crack like a stockwhip. Then, suddenly, it broke. Within a few hours, despite the immensity of the downpour, there was a bloke from Telstra, standing on the far bank, using a fishing rod, of all things, to try and flick a line across. When he miraculously succeeded, we tugged at it and, in due course, could haul a replacement cable into place. Those were the good old days for Telstra, when people in the rural areas got something resembling service and Telstra blokes were local heroes. Little wonder that people in the bush are so hostile to privatisation.

Why isn't the river as narrow as it should be? It's because of timber cutting. The removal of trees has encouraged sediments to move from one side of the river to the other, a process that has encouraged the widening of the flow and the erosion of all outer bends. One bend has been threatening a house, worth perhaps $300 000, for years and the DLWC has spent hundreds of thousands of dollars (some say two million) trying to stabilise the bank and save the house. Many suggest DLWC should have bought the house twenty years ago and saved their budget instead.

In the past, small floods would have entered the floodplains comparatively gently, releasing their sediments. The Rivercare plan notes the river needs to be allowed to flood, but who'd want it now that the area is so developed?

All these efforts to 'improve' and 'save' the river are like 'using a bandage to repair a broken leg,' says one scientist. Should the Pages be proud of being the 'most worked on' stream in the Upper Hunter? Because the river is so 'stressed' and such a 'high energy stream', no less than 106 works have been carried out over its 83 kilometres since the 1950s – unlike the paltry 68 works along the 265 kilometres of the mighty Hunter River that flows above Muswellbrook. How sad that 75 per cent of the repair work involves repairing the repairs. Floods have a bad habit of wrecking rehabilitation.

Nonetheless, old photos reveal Segenhoe was in a worse state back in the 1950s, when dairies proliferated and tree cover was negligible. It's to be hoped the latest rock revetment will be a lasting answer to collapsing banks.

The Hunter catchment now has a new blueprint, which will be implemented over the next ten years. This document attempts to incorporate all the ideas, all the knowledge, about soils, revegetation and river flow – to encompass the whole ecosystem of the Hunter Valley. Yet its implementation can succeed only if we all understand and support it.

Which brings me to a multi-million-dollar revegetation project downstream on the Hunter River. South of Muswellbrook there's a bridge called Keys (named after the original white settler) that takes you across the Hunter River to Bengalla Coal Mine. It's the start of an 8-kilometre project that will see willows removed, snags reinstated, and native long-stemmed trees planted. A big, costly and impressive experiment in ecosystem management, its progress will be monitored by scientists to establish benchmarks for the future.

Research work is, of course, of fundamental importance; and, while I love hearing about it, I don't like learning that two coal mines get a 125 per cent tax deduction for their $250 000 donations to the project. In other words, the government pays (overpays) for their contribution – and the mines get to boast of their services to the environment in PR brochures. The only way farmers along the Pages can get comparable funding is in the unlikely event of a scientist approaching them to set up a research project.

I want to fence off one section of our river and am shocked when I get a quote for $13 000. Why is it so expensive? It's only 1.4 kilometres long. The contractor explains that it's all up and down, with eight deep gullies to cross, requiring strainer posts, stays and netting and lots of extra labour. So I investigate if any subsidies are available, and discover that there's some left in the Telstra kitty – equivalent to $2 for every kilometre of fencing. Compared to the mines' 125 per cent tax write-off, it's a joke.

It was Peabody Resources Ltd (part of the Peabody Group USA, the world's largest coalmining company, producing over 150 million tonnes per year) that set up the joint venture to develop Bengalla Coal Mine. They claimed an investment of half a billion dollars and published glossy brochures to talk up their 'commitment to the Hunter Valley and community' as well as to the coal seams. (Corporate self-aggrandisement is, of course, directly connected to America's 'self-improvement' ethos. You're taught to sell yourself, even your soul, in the interests of success. And as success remains ever elusive you're destined to forever climb a mountain without a peak.) In any case, I've learnt that it isn't a good idea to believe what people and companies *say* they are. They are what they *do*.

An early Peabody brochure promised to 'build upon history', 'minimise noise', 'monitor continuously', 'plant trees', 'manage water', 'listen to our stakeholders', 'talk to each other' and 'milk cows and stable horses'. Far from being in the business of digging coal, they're building a stairway to Paradise! So, in 1990 they won the right to explore for coal in the area. In 1999 they loaded their first shipments.

And in 2001? They sold out – so much for their long-term commitment – to Coal and Allied, a mighty enterprise 70 per cent owned by Rio Tinto. In 2002, $224 million was paid in wages and $12 million in royalties – because for every tonne of coal sold, $2.60 goes to the state government.

A mere 4 kilometres from the centre of Muswellbrook, Bengalla Coal Mine is surrounded by farm land that, over the years, has been subdivided to accommodate dairies, lucerne farms and rural retreats for horse lovers. I stand high on the overburden hill and look down over a blissful panorama, with the Hunter River meandering through it, on to the agricultural businesses owned and managed by Bengalla – in line with the undertaking to 'milk cows and stable horses'. But not everyone is convinced of its good intentions, arguing that this has always been window-dressing and that Bengalla's strategy is to buy out and operate farms as a way of reducing the level of complaints.

As you drive up to the mine, the whole area is surrounded by crops and Friesians chewing the cud. It doesn't actually look like a coalmine until you're right in it, dwarfed by its drama, scale and sense of power. The most noticeable thing is the lack of people. When you look at the mine from the highway, it's rare to glimpse even a truck lugging overburden up the hillside. The mine seems to mine itself. Nor is it particularly noisy. As we drive around the site, the only sound is the wind.

Yet ahead of us is a $65 million dragline gouging the metres of dirt on top of the coal, its mighty shovel forming a giant heap that will, in due course, create a small mountain. The word from the town is that the new Twin Towers are being built at Bengalla. As the piles rise to the heavens, they're constantly reshaped, sweepingly concave one day and like pyramids the next. It's a remarkable site/sight. Far more a work of art – the sort of thing Christo might undertake – than anything resembling the natural world. But most people in Muswellbrook don't want an immense sculpture. They want, one day, for everything to look the way it used to.

At the hoopla that inaugurated the venture, politicians of all denominations shared the ribbon-cutting, the speech-making and the champagne. Some 180 people gained full-time employment. Yet, despite all this, there remains an underlying diffidence about the project and a tendency to camouflage rather than proclaim. Though there's a well-attended Open Day each spring, the company limits its community relations to the controlled media of glossy brochures and press releases, as if Bengalla is just a little anxious to maintain its image. As a parent, I wonder how the mining executives feel when their children come home from school and nag them about global warming.

For it has to be admitted that burning coal for electricity is a fossil industry, in two senses. It burns geological fossils to fuel our up-to-the-minute lives; and, by playing such a major part in global warming and creating unacceptable levels of carbon dioxide emissions, it is an industry doomed to become a fossil. It's been said: the Stone Age didn't end when they ran out of stones, and the Bronze Age didn't end when they ran out of bronze. They ended when human beings found a better way. The Coal Age should have ended decades ago.

Should we feel proud of an industry that gives short-term gains to some in a local community and to industry shareholders, while simultaneously endangering the planet? No wonder Bengalla built high walls to hide their conveyor belts. It would be better, more honest, to erect giant neon signs on the overburden hills blinking out messages like: 'YOUR HOUSE IS MORE IMPORTANT THAN GREENHOUSE' or 'TO HELL WITH THE FUTURE.' It would look like a Michael Leunig cartoon, and would, in fact, be far more truthful.

So what price our river – our little river? It is hardly the Murray. It isn't even the Franklin. Bickham Coal Mine is hardly Bengalla. Unless, of course, someone is keeping something from us. Will anyone care? Will the Greens? We tell Bob Brown, hero of the Franklin, about our coalmine dramas and a few weeks later he arrives, immediately dunks himself in the Pages, sets about meeting people and tries to visit the Bickham site. They're disinclined to unlock the gates. Although almost 60, Brown has the enthusiasm of a teenager and a simplicity and directness that impresses most locals – though not our state MP, George Souris, then head of the New South Wales National Party. He goes on radio the day after Brown's visit suggesting the Pages River and Bickham Coal Mine were none of Brown's business and that we didn't need 'philosophical extremists' in the district. Bob 'Bin Laden' Brown laughs. He's heard it all before. Lee Rhiannon, from the New South Wales Legislative Council, picks up the issue.

If our enemies had their way, they'd be the only ones allowed to talk about anything. Gagging debate, as George Souris tried to do, is the main tactic. Now we're advised by Bickham Coal that they'll 'agree to disagree' with us. In other words, it doesn't matter what we say, the mine will proceed.

It's been a long day. Latching the gate of the chook pen, I head back to the house, detouring to the dangling purple muscat grapes that shade the asparagus bed. There's so many I'm profligate. Standing under the vine, I greedily pick fistfuls and shove them into my mouth for a sweet explosion, a sort of pre-dinner splurge. The best bunches hide behind leaves, protected from birds and me. I reach for the deepest, most unspoilt cluster, tearing off the grapes. Reaching up, I notice someone has picked at that bunch before me and, moving to the next perfect cluster, I brush my hand against a wasp nest. It's beautifully made, with the same hexagonal pattern as a beehive, and the colour of spent autumn leaves. It is delicate, spherical and lantern-like, a sculpture the size of a small apple.

But the slight touch of my hand is a crisis for the wasps and, in an instant, hundreds are attacking me. No sooner do I realise what I've done than I'm stung all over my hand, arm, neck and face and run screaming to the garage to grab the antihistamine spray kept on hand for emergencies with bees. *Squirt! Squirt! Squirt! Squirt!* Then I rush to the swimming pool, strip off and plunge in, all the time thinking of the stories about bites on throats or tongues that swell and block windpipes.

I'm used to bee stings, but nothing prepares you for a wasp's. Even the cold water doesn't help. When I climb out, Aurora counts eighteen red, throbbing itchy welts.

Most people who've worked here have had an altercation with wasps. Welts can double in size. Even lads in their twenties, despite all efforts at bravery, want to, and sometimes do, cry.

There are so many different wasps around here: all natives, all doing their bit gobbling bugs. Nurserymen preferring biological controls to insecticides want them attached to their glasshouses to provide an ecological balance, but wasps attach their paper or mud homes to all sorts of inconvenient places – like the inside of a raincoat left too long on the peg, or behind the stereo in the kitchen, above doorways or hidden dangerously in olive trees.

The largest and most ominous-looking wasps are hornets, completely harmless to humans. They build muddy domes all over the house, even in our hats. We remove and demolish them, revealing white pupae inside convoluted tunnels.

Along the river, wasps form colonies and cities in caves. Once, thinking I was following a bee track, I ended up standing surrounded by them. It was like being in the middle of a minefield. Had I accidentally touched one? Getting too close can spark a change of mood and the pitch of their buzzing rises and they dart anxiously: a thousand little De Niros from *Taxi Driver*, 'You lookin' at me?'

When the lowering sun begins to soothingly filter the light, and the dried-up trees don't look so withered any more, instinctively I begin to look for mushrooms, knowing they'll appear after the slightest rainfall or heavy dew.

Years ago we'd heaped left-over lucerne silage into big mounds for compost. Six months later we shovelled the rich, rotted result all over the vegie garden and, within days, Swiss Brown mushrooms erupted. Obviously the lucerne compost and the Swiss Brown are somehow married. If I spread the compost around the violets or under roses, there's a mushroom display.

I never buy mushrooms, because the paddocks have always provided plenty of field mushrooms (*Agaricus campestris*). Phillip and Aurora head off, singing their silly hunting song, 'What do we want?' 'MUSHROOMS!' 'What don't we want?' 'TOADSTOOLS!'

At first, excited by a new variety and flavour, I reshuffle the menu to include as many as possible. Stroganoff, mushroom soup. Coq au vin becomes mushrooms with chicken. While Phillip could have them sautéed on toast every morning with a heavy sprinkling of parsley, I enjoy them best slow-cooked in sherry, with a dollop of sour cream, a sprinkle of parsley and pepper and wrapped into filo pastry parcels. After two weeks though, you've had enough and want them to remain underground. Even friends say, 'No more, thanks.'

There's nothing quite like a mushroom harvest to help you focus on all the inedible fungi that colonise many a rotting tree and damp crevice. There they can stay, highlighting a walk. Take, for example, the spectacularly bright beefsteak variety, which has orange, semicircular saucers jutting out from tree trunks like wasp-stung lips. When we take a rest and sit on a log, we're just as likely to discover next to our bottoms a fungi less showy but equally remarkable, the white gelatinous 'jelly fungus', all rubbery and gathered like a skirt. Or delicate caps clustering on a broken branch. On some autumn days there are so many varieties that a brief walk in a gully leads to cries of 'oh look' a dozen times over. They'll share the season with numerous ferns that have died back in summer, hidden under tall grasses only to reappear, uncurling from the leaf litter.

Maidenhair, the most delicate of all the ferns growing here, will clump together, highlighting an entire east-facing rocky hillside with glistening fronds. Along the Pages, where the rocky outcrops curve in and around the river's edge to announce a hidden gully, you'll find maidenhair clinging to the fractures in the rock face, surviving on dew. And below, as if part of the family ('You guard the rocks; I'll guard the water'), floats the watercress, ready to spice any salad.

There's good news. Tests show that our water isn't saline and that the quality of clay in the gully is promising. We're ready to proceed with our dam. Days later Col arrives with Terry and they unload a great grumbling, grunting dozer from the back of a truck. Within minutes the drama is under way, with Terry at the wheel, and the gully is massively reshaped. More anxious to collect water than ancient artefacts, Phillip phones the farm every few hours asking for a progress report.

I'd hoped to run a new, bigger pipe from the river's pump to the new dam, so that it could be quickly filled and topped up whenever the Pages permitted, but it turns out that this would cost more than the dam. So we're going to keep using the old narrow pipe, pumping water up to the tank on the top of the hill, and then gravity-feeding it to the dam that Terry is boldly, confidently creating.

And within a few days, he's done it. A great concavity that will, fingers crossed, go some distance towards ensuring there'll always be enough water for our olive grove. The curved wall of the dam is very impressive – and Aurora runs laughing up and down it.

Before opening a valve at the tank to admit the first trickle of water, we surround the dam with a single-strand electric fence to keep the cattle out. This water will have to be pristine, unmuddied. At the 'opening ceremony' the water rushes downhill through a black plastic pipe uncoiled for the purpose. The flow doesn't look much more than you get from a garden hose. So whilst the river is running high, we keep the pump thumping away, day and night. Phillip pushes sticks into the clay so that we can calculate the rise in water level; and for the first couple of weeks, we're mesmerised as they're progressively submerged.

He and Aurora christen the dam with a clear, cold swim, and we wonder when the first turtles will arrive. However, growing excitement becomes anxiety when we realise that the dam's not getting any deeper. Phillip's sticks prove it. Is there a leak? We turn off the pump to check and, overnight, the dam sinks and sinks and sinks. Phillip and Aurora go exploring and find that it's leaking into a creekbed 40 metres away – a clear trickle into a flurry of watercress.

'It's leaking!' thunders Phillip, and roars about waste: the waste of electricity; the waste of pumping time; the utter waste of water. Not to mention the waste of money on the dam itself.

As he gets angrier, I try to remain calm, phoning Col to describe the problem. 'You'll have to let the water go,' he says. 'And then we'll come back and redo it.'

So every day we watch the water disappear from the dam and reappear, through some mysterious crack we can't locate, in the creek – eventually, of course, to find its way back into the river.

The art and science of dam building has failed us. All the soil and water tests were positive and no one handles a dozer better than Terry. Col is both embarrassed and mystified. 'I don't get it,' he says, shaking his head.

While Col worries and Phillip rages, I wander down to the river to look at the fallen tree. The rains last week brought down a small wave of water and, for a moment, the trunk was submerged. But it didn't budge. Now the river is flowing freely, shallow enough to reveal the rocks on the bed and the carp and catfish swimming above them. Despite its long journey, the water that began flowing by the Capuchin hermitage is bright and unsullied. The shimmer of the water is enough to make me feel washed and cleansed.

How my opinion of the she-oaks has changed. At first I saw them as desperate-looking and untidy. Many produce a brown flowering in February that does little to enhance their beauty or that of the landscape. But, gradually, I've come to admire their adaptability, their tenacity, the way they deal with drought. And I've accepted their appearance, seeing them as an Australian counterpart to – and as defining of the landscape as – pencil pines in Italy.

The one that lies dead in the water had broadened with age. As well as its roots sticking up, one branch protrudes, its brown needles waving to me. The change to the riverscape wouldn't matter if the river were left to its own devices. But I want to pump here; I'm making demands upon it; the river has to fit in with my plans.

Folders on my desk contain a glut of information about the river. There are analyses of its woes and proposals for its redemption. After nearly a couple of centuries of indifference, we are now diagnosing it to death. There's a chance the river will benefit from all the studies and bureaucratic examinations – that it may be permitted to flow at least a little freer.

Many, who resent the new proposals and are sensitive to any criticism of past practices, maintain they've nothing to apologise for. 'The river has enabled five generations to make a living. We've made export dollars for the country.' Certainly the river has helped support communities, both blackfella and whitefella, over thousands of years. But now the claims made upon the river, and for it, are intensifying and conflicting.

'If I can't irrigate,' says a neighbour, 'I can't have the type of place I want.'

What's easier? To change the irrigation rules? Or to change human desires? I know the answer. Desires are insatiable, which is why we have rules.

Our first attempt to live with the new rules was to build the dam. But it has failed. Our dam, so nobly designed, was going to fill when there was water aplenty. It would stop us yielding to the temptation to pump when the flow was low. We'd be able to boast of planning in advance, of shielding ourselves from drought, which is what urban critics demand of us farmers. Instead, our dam is not merely a disappointment, it is a disaster: an ugly crater that is filling, not with water, but thistles.

Yet I won't give up. Soon Col and Terry will be back and, somehow, they'll make it work. All we need is time and patience. So I sit on the western side of the river as the sun leaves the stage;

shadows, slowly falling over me, try to cloud my thoughts, but I resist. The balance we're seeking must be attainable, achievable. There is simply no option.

Back in the kitchen the flashing light of the answering machine demands attention. The builder has phoned, confirming the arrival of two men who, tomorrow morning, will pull the kitchen ceiling down. After enduring sixteen years of bad lighting – meaning that the three of us need three separate lamps when we're reading at the table – I've decided I can't take it any more. Elmswood's ancient wiring and the cracked ceiling just won't endure more patched-up improvements. So tonight we have to move most of the kitchen gear into the dining room.

'But how are you going to cook? What are we going to eat?' asks Aurora, alarmed at the disruption.

I dust off the jaffle-maker.

Private irrigation schemes aren't new to the area. I discover that the first one ever proposed in New South Wales was way back in 1890 on the Pages in Segenhoe Valley. The Land Company of Australasia wanted to dam the Pages and Rouchel Brook to the east. Local residents angrily protested and the government of the day was suspicious of a large company grabbing water rights. In 1893, after failing to irrigate their 25 000-acre Segenhoe Estate, the company was declared bankrupt and the land subdivided and sold.

There's a couple of private irrigation schemes operating in the Lower Hunter, where water stored in Lake Glenbawn is released and channelled via a network of pipes to contracted thirsty vineyards and homes that are still benefitting from increased property values.

When Scone Council began planning the shire's future water requirements, a far bigger pipe from Lake Glenbawn was proposed. Why not have a private scheme tap into it, so that people other than townsfolk can also have the bonus of a reticulated supply? The water could be spread out along any road, as long as there were enough people willing to sign up and pay for the service. In some ways the proposed system would parallel a co-operative.

Of the beautiful Lake Glenbawn's vast holdings, 95 per cent is earmarked for Macquarie Generation for their power stations – not for hydro-electricity, but for cooling the giant turbines. Only the remaining 5 per cent is available for irrigation. It has been suggested that if the level of the dam was raised just half a centimetre, this would provide enough water for all of Scone Shire – which wants a mere 11 000 megalitres, plus 4000 megalitres for a private scheme.

The thought of being able to extract water from a pipe rather than the poor old Pages might make all the difference to the Pages, as well as us. Or would it simply mean more unsustainable development?

Elmswood's large licence to pump from the Pages isn't a lot of use when the overall increase in upstream irrigation is sending the river into imposed droughts more often than ever. Here is a supply that could keep our olives well watered over summer, so we sign up for the scheme and start thinking about another holding dam.

The hardest thing for farmers to accept is the limitations of their land. Unlike other businesspeople who can make adjustments when they need to 'grow' their business by adding staff and machines, farmers are locked into 'developing' their land. And, of course, 'developing' is a dangerous word. Often it means doing things to the land – such as over-fertilising, clearing, poisoning native grasses to plant greedy, introduced species – that provide short-term gains with detrimental, even catastrophic, long-term effects. But the first and foremost input needed is water. Every development in our shire, in Australia for that matter, is dependent on water – and 'water reform' hasn't really made water use more efficient yet. It has, instead, largely directed water to those who have the most money.

The day after our Water Association meeting, when it was painfully clear that none of us had the foggiest idea how much water we were really using, I booked into the WaterWise course, a free service provided by New South Wales Agriculture with help from the DLWC. I'd been procrastinating about taking it for many months. Now it was a matter of urgency. A four-day course spread out over a few weeks takes place in a winery in Pokolbin. Amongst my fellow students are vignerons, lucerne-growers, dairymen and women, and olive growers. Once again, I'm provided with folders bulging with pamphlets and information, with documents headed 'Scheduling', 'Benchmarking', and 'Learning Outcomes'.

Soon we were trying to calculate how much water is irrigating a particular crop, a simple procedure involving basic arithmetic – simple, as long as you know the texture of your soil. To be efficient with water, you have to understand when the saturation point has been reached – or to use the technical term, the 'field capacity.' If you continue to irrigate past the point where the plant accepts the 'Readily Available Water' (RAW), you're wasting it. Over the next few meetings, learning the RAW will be crucial.

'You don't want to push past the field capacity,' says Norm, our instructor. 'That's when plants start to use energy to absorb the water.'

Since different soils have different RAW levels, knowing about soil texture is paramount. So we stand in pits and rub soil between our fingers, feeling the subtleties of sand and grittiness, of clay and stickiness. We measure water-infiltration rates. We evaluate sprinkler performance. We turn data into graphs. And all the time I'm feeling appalled I don't already know this. Pretty soon, I'm calculating the Mean Application Rate (MAR) with the Distribution Coefficient (DC). As for working out how long to irrigate – that's simple. It's simply the RAW divided by the MAR.

What I thought I already knew was how much water we'd put on our olive trees. I'd based this on the amount that our irrigation system was meant to supply. If I left it on for an hour, I'd given each tree four litres of water. But when I get home and run a few tests, I discover that I'm in error.

This, I kid you not, is the state-of-the-art measuring technique: you take ten takeaway containers from the Chinese restaurant in Scone and place them at various distances from the sprinkler head. You turn the sprinkler on for a set time and measure each container. From observing what was captured by the short-soup containers, the fried-rice containers and the steamed-vegetables containers you can confirm whether the water is reaching where it's meant to reach – and if so, how much and over what time.

To our surprise, we discover that we've been putting far less water on our olives than expected. Less than the sprinkler and drip manufacturers suggested we should be getting. We also discover that the watering wasn't even across the grove. The variation wasn't huge (8 per cent), but that can add up to a lot of water, or a lot less water, during a dry season.

The problem seems to stem from the main pipe from the tank to the filter. Even when the tank was brimming with maximum pressure and when we'd run the system with all the air valves open to ensure there were no blockages, we weren't getting the quantity of water we'd expected. There was less water coming from the system than was the case back in 1997 when we'd installed it. One theory is that frogs get in the pipes and die, stopping the flow, or that the main pipe rises and falls in the black basalt, creating air blocks. You can only discover problems like this if you dig up the whole system to investigate.

I then place the takeaway containers in the lucerne paddock and discover problems there as well. Over the years, the umpteen sprinkler heads on the pipes have been damaged, worn, or changed with heads of slightly different design. So without noticing, we have been overwatering some parts of the paddocks, whilst neglecting others.

'Mum,' says Aurora, 'can we have Chinese food tonight?'

'No, I'm cooking risotto.'

'But don't you need more Chinese containers?'

8

Dramas notwithstanding, two-thirds of the trees in our first grove have survived. Despite frosts, lace bugs, cockatoos, roos, pigs, scale, harsh pruning and gale-force winds hammering at their blossoms, we still have little olive jewels dangling from some branches and feel certain we'll get our first bottles of oil this year.

Time has taught me an invaluable skill: how to protect myself from disappointment. When it comes to life on the farm, whether it involves rainfall or cattle prices or making hay, expect the worst. So it is with our first olives.

Each morning before the heat of the day takes hold, Phillip and I water a section of the main grove and check the trees for ailments. Up and down the rows we trudge. For a month, there's nothing untoward. The pumps suck, the tanks fill, the filters filter and the olives drink, albeit unevenly in the summer heat.

We leave the 1000-tree dryland grove to fend for itself. Not out of unconcern or benign neglect, but because there are just not enough hours in the day. Moreover the dryland grove remains my great experiment. If the trees are strong to start with and well watered when planted, they should, so the theory goes, bear fruit for centuries.

But in the main grove, by the homestead, seeing olive fruit on young trees, which require just a tippy-toe stretch to reach, gives a sense of everything being within our grasp. Not just the olives, not just the trees, but the whole idea of the enterprise.

Anzac Day, 2002. After the march in Gundy, attended by veterans from all over the district, and after the observations at the Gundy Memorial Hall, where this year I had the honour of speaking about Gallipoli and the war, we have a ceremonial occasion in our olive grove. Two of our best friends, Christine and Peter, recently returned from living in London, arrive to help. They've been warned that picking will take at least two days, but the olives are few and far between. In truth, only about one tree in ten has much to offer, so it will probably all be done in a few hours. The trees are full of surprises. Some of the largest and strongest have produced nothing, whilst smaller, spindlier trees are bountiful.

Olives ain't olives. Leccino, Correggiola, Frantoio, Bouquettier and Manzanilla are all different in size, in weight, in content. Pickers like Manzanilla because they're big and plump and fill a basket quickly – but they weigh considerably less, being full of water rather than oil.

I find one aspect of the pick humbling. Despite practising, I still lack the ability to stand back from an olive tree and assess the quantity. All growers need to be able to do this – how else can they plan their processing schedule? But again and again the scales prove me wrong.

Aurora and her friend Molly are helping and triumphantly carry their full baskets to Phillip, who loads them on the truck. There's clearly competition amongst the grown-ups to see who can fill their baskets quickest. At the same time, there's a point of pride in making sure that each tree is picked clean, that precious fruit isn't left behind in the rush to the winning post.

This first harvest, the harvest that since 1997 seemed increasingly improbable, has been a modest success. Mind you, we were just in time. For two years I'd eagerly anticipated the olives turning black. What month? What week? What day was likely to be the best to harvest? We'd conducted training exercises, in 1999 harvesting wild olives from an abandoned grove we'd found on a small farm 20 kilometres away. But we'd picked too early, so the percentage of oil extracted was abysmal. On the other hand, the rough, green and pungent product lasted three years in the cellar.

We've waited until every olive has turned black. Some of the Correggiola trees on the top of the hill have already lost olives to over-ripeness or birds, and we find telltale pips and shrivelled fruit scattered about. Squeezing the soft olives in our fingers we rub the oil straight onto our faces.

Resting on a blanket in the grove, we eat a snack from Christine's picnic basket, then drive the loaded truck down to John's farm in Pokolbin for processing. John is a 747 pilot with an international airline. We met years ago when he came to buy a donkey as a surprise present for his wife, and left with two. One of them had been improbably christened Venus Toboggan by three-year-old Aurora – who is more excited about the thought of a reunion than about our harvest.

John had begun with a vineyard and was now planting olives. Then he lashed out on a state-of-the-art press and was spending his spare time pressing neighbours' fruit. Aurora throws her arms around Venus, while we present our modest crop to John, who is clearly pleased by it – not by the quantity, but by its ripeness. Nothing is worse for a processor, for finely-tuned machines, than trying to squash oil out of unripe fruit. Oil pours from ours.

The noise of the processor is as deafening as one of John's jumbos and, though we are not flying, our spirits soar. Elmswood's first oil looks wonderful, miraculous. As the fruit is crushed and mixed, we watch as the oil begins to ooze from the mush. The smell is peppery, even sexy.

When Alexander the Great died, his body was preserved in honey and sent back to Babylon. Today, I decide I want to be preserved in olive oil. The aroma of the oil is more than enough to compensate for the numbing drumming of the machine that produces it. There is only one disappointment. The oil isn't green enough. We'd let the olives ripen to such an extent that it is a pale yellow-green. Yet everyone who tastes it thinks it's fine: fresh, clean, aromatic of fruit and flowers rather than grass and hay. But is it the type of oil I want to make? I don't think so. I still hadn't found quite the right moment to harvest.

The following year, 2003, our olive grove produces nearly

twenty times the fruit. The drought hasn't hampered the crop and the trees seem to be thriving under stressed conditions. Gavin, our new manager, and our new young helper in the grove, Colin, start the picking, and Christine and Peter return, this time with their three grown-up kids and some friends. For three days we harvest the fruit, taking truckloads down to John at sunset.

The trees are bigger, and we have to park a truck alongside them to reach the tops. I walk the grove in my bee-suit and cut out three wasp nests, but the trees have other creepy-crawlies like golden orb spiders. Christine warns me that daughter Isabella is particularly arachnophobic, having had a close encounter with a huntsman lurking on her undies as a child. There are loads of them as well. One lives in the truck and there are always some in the bedrooms.

There is simply nothing I can do about the orb weavers so I tell Aurora seriously, emphatically, not to mention them and she looks at me as if I'm mad. As if no one will notice them! They are, after all, everywhere in the grove and, as far as spiders go, pretty big, with bodies exactly the size and shape of olives. Some defiantly display their spectacular golden webs between trees cluttered with trapped muck, while others tuck their legs under their body and hide, just waiting to be mistaken for fruit. The whole grove is alive with bugs and their activities. For instance, wasps have been observed landing on an orb's web, imitating deathly vibrations to lure out a spider in order to paralyse it and store it as live food for its young.

Another tree has a twiggy pigeon nest with two white eggs in it, and a mother who keeps guard by fluttering noisily back and forth. We pick carefully around her nest. Two weeks later we visit a contented nestling.

Everything I ever wanted for the olive grove has come true. And the harvest is made all the more enjoyable by having friends wanting to share it and enjoy it. Peter and Christine have to be the most generous people we've ever known; generous in every way – with their time, their spirit and their sense of humour. By the end of our harvest, I'm not sure if what really excites me is the perfect oil streaming into the stainless-steel vats or the experience of their friendship. Either way, I feel that if our grove were ripped up tomorrow, like those in the contested hills of Israel, the memory of our first 'commercial' harvest will last us a lifetime.

I like the comfort of familiar birdsong. Before dawn the rooster's crow seeps into that 'I'm becoming conscious' moment, reassuring me that I'm home. After a fox killed one rooster, a speckled ancona, the mornings became eerily quiet, leaving me feeling odd all day. I slept uneasily and couldn't wait to buy a replacement.

I want birds to nest close to home. Many do. Outside the kitchen there are willy-wagtails. My favourite climbing rose, Albertine, houses double-barred finches; magpie-larks make mud nests in the box tree near the chook pen; magpies build twiggy nests in an ironbark next to the barn; and pigeons have claimed many an olive tree as home.

Occasionally the nests are destroyed, or the eggs taken by

currawongs and butcherbirds. There can never be enough small birds: the insect- and seed-eaters, with their light melodic chirping. Despite the initial annoyance, even the endless tat-tat-tat at the window by a delinquent rufous whistler is fascinating to us and, apparently, to its mate, who watches and listens from a perch on a nearby branch.

Once we found a very high-class racing pigeon pacing around the garage. Within days it was feeding out of our hand. Lost? Confused? Hurt? We hoped it would stay, but then, just as we were making friends, it flew off. We had been hearing about GPS devices and mobile phones, with their electromagnetic waves, making the skies dangerous places for birdlife. The British Trust for Ornithology announced the common sparrow count was down 50 per cent, correlating this to the introduction of mobile phones. There's a concern that the radiation might affect their reproduction. Around here, the double-barred finch count seems down, and I'm wondering if the GPS I use when mapping the fence lines is having a detrimental effect. Is a 'silent spring' in the making?

Thinking about what's missing can become a pastime in the drought.

9

By spring 2002, the river isn't a river any more. It's not even a creek. You couldn't even describe it as a chain of pools because the links have been broken. If there's any flow, it's underground and out of sight, disappearing into what's called the hyporheic zone, the scientific name for 'below the riverbed.' In between these sand and gravel layers is a dynamic ecosystem where bacteria and invertebrates live, all doing their bit to keep the river healthy. Research into the Pages' hyporheic zone has only just begun. So far, the data is fascinating. There are at least thirty-two invertebrate groups and several mite species that may be new to science. It's been suggested the Pages could also be the breeding ground for many invertebrates heading off downstream into the Hunter. Since Lake Glenbawn stops the migration of many invertebrates, the Pages' contribution to the Hunter's biodiversity is important.

The hyporheic zone is never static. Every flood pushes downstream not just the rocks and gravel on the top, but also the particles a metre below. These pebbles and rocks I'm walking over now have been here only since the last flood.

Evaporation leaves a scum across the riverbed that is baked by the sun into a glistening white, ceramic coating. Bright waving weeds are colonising the riverbed, poke root (*Phytolacca octandra*) and smartweed (*Persicaria decipiens*) are popping up in damp crevices. At a glance, the riverbed is a dull grey; on closer inspection, there is great variation and colour.

It's impossible to walk here and not pick up pebbles and rocks. Every single one strikes me as beautiful. So over the years, hundreds have found their way up to the house to be used as doorstops, paperweights, table decorations. Pebbles chosen for their distinctive shape or colour form rock mulch around geraniums, whilst others fill pots, alongside Aurora's piles of seashells, gathered from every beach she's ever visited. Extra special rocks – chosen by my Mum because they were polychromatic or strangely shaped and which she arranged whimsically around Elmswood – are now displayed on my office mantelpiece upstairs.

They are all symbols of geological time, what Darwinian scholars call 'deep time.' Created and shaped by unimaginable processes and immense forces in volcanic furnaces or in gravity's presses, they are the fruits of the river, made easier to collect now by its disappearance.

Ever since I've been with Phillip – and it will be twenty years soon – I've been surrounded by rocks and stone artefacts of all sorts. Near me, as I write, are a handful of perfectly spherical stones long ago employed as projectiles in the sort of slings that David used to slay Goliath. There's a piece of broken ancient Greek stone shaped before Christ, and some fragments of white limestone from the quarry used to provide the stones that once covered the Great Pyramid of Giza. There's a ventifact Barry Jones gave Phillip on his return from Antarctica – collected from a bone-dry desert where neither rain nor snow ever falls and where every rock is shaped by wind. It's as black as ebony and very dense, yet wind and wind alone has sculpted it into what looks like a piece of aircraft wing. It's curved on one edge and knife-sharp on the other. Imagine something as soft as air doing this to something as hard as rock.

Then there are the statues Phillip has collected from virtually every known and forgotten civilisation. A huge piece of black basalt is carved into the head of a baboon, one of the creatures that used to dance on the rocks at the Nile's first cataract. From a piece of grey stone a Khmer artisan has created a calm image of the Buddha. There are Roman figures from Carrara, the quarry that's still being used today and where Michelangelo found the giant block of marble in which he saw the form of his *Pietà*, now displayed in St Peter's Basilica. And yesterday Aurora brought home a stone simply because it felt nice to hold.

Every time in deep time, every time in human history, has been and is a Stone Age.

I gather my latest collection from the dry riverbed to show geologist John Roberts, Emeritus Professor of Geology at UNSW. John has walked many of the creek beds and hills of the Upper Hunter. 'I try to walk from the oldest to the youngest, but there are times it has to be the reverse. Meaning my thoughts have to run in both directions.'

Looking into my latest box of rocks, he says, 'You hardly ever know where these things come from.' As I number each one with a felt-tip pen, John scrutinises them through a powerful magnifying glass. There's quartz of different shapes and types, ignimbrites coloured by the iron-stained zeolite minerals.

'This one,' says John, 'is similar to what I called Elmswood ignimbrite that I found years ago at the foot of Black Mountain.'

I like it, not just because it's named after our place, but because ignimbrite is formed from the crystals, pumices and ash deposited by the pyroclastic flow in the Carboniferous Period. In other words, the ingredients that made this rock blew out from a volcano 354 to 290 million years ago, probably farther west, at immense speed.

John also identifies rhyolites, coarse-grained sandstone, pebble conglomerates, grey silty mudstones, Permian basalts and a pink granite identical in hue to the stone used in a larger-than-life portrait of a Middle Kingdom pharaoh that looms over us in our hallway and was sculpted 3000 years ago near Karnak. Is my pebble from the Pages related to this stone from the Nile? A green siltstone, collected from Cameron's Gorge, is like nothing we find at home. Immensely hard, its grain is so fine that it fractures with a concavity. Our siltstones are different in colour, but also composed of the finest of materials from the same Carboniferous Period.

John spreads out four large maps he's coded with soft colour pencils, highlighting the eight different geological formations the Pages River passes through before reaching us at Elmswood. The receding river reveals proof of our region's 350-million-year-long history. The stones take us back to the Carboniferous Period, a time before the coal seams of the Hunter Valley were laid down, a time when the landscape would have been as voluptuous and volcanic as Indonesia's and as cold as the southern tip of South America.

'The bitumen-coloured basalt rocks at the head of the river are the youngest,' John says. 'These are Tertiary, around 40 million years old.' It's the earlier times of the Carboniferous that have long occupied John's geological interest, so he's dismissive of the younger Permian and Triassic times. 'There are no Carboniferous basalts in this part of the world.' I take hurried notes, trying to keep up with his impromptu lecture. I learn that, at the hermitage, the river is young and vigorous. As it leaves Segenhoe Valley and disappears into the Hunter, it's old. The flow of the water is, in a strange way, a flow through time.

I find it difficult to grasp, but the river is younger than the Carboniferous Period. Since the riverbed is the lowest point on the property, there's a sense of its depth signifying deep time. Yet it turns out this is an illusion, a confusion on my part. The pebbles I'm now collecting from the riverbed come from the youngest part of the property. The banks beside it and the cliffs that can soar 20 metres above me are much older. Everything is eroding, grinding away. Every collision of every pebble, every abrasion of larger rocks, is driving the river and the surrounding land to its future flatness. Slowly and surely, from pebbles to sand grains, the edges are being knocked off the old.

This is what has happened to the mountains around us. Once upon a deep time, Elmswood was almost Andean in scale.

There's clear evidence of volcanoes all around us. Now, looking out the kitchen window, I can imagine it. Majestic, snow-covered and thundering: Mount Fuji in the garden.

The river is being choked with sludge. The dreaded green weed – a filamentous alga of the *Cladophora* genus – clings to rocks and foot valves. The mere act of looking at it seems to encourage it. It spreads as inexorably as a rumour, as a contagion. Or perhaps all the pumping for lucerne and pasture is simply revealing it, like an ill-kept secret.

The weed is a light, bright green that would appeal to fashion designers. When the river was a little higher some attached to willow and she-oak branches, forming long, leather-like strips. Rotting piles provide curtained boudoirs for moorhens and teal ducks.

Preposterously prolific, it looks nutritious. Is there some way to harvest it? Could we drag it out of the river and feed it to the cattle? Phillip and Aurora experiment, pulling long skeins of the wretched stuff, rolling it up in bundles and dumping it in a paddock. No creature attempts to eat it. It turns out to have the texture and toughness of fibreglass.

Floods are expensive. They destroy fencing, drown crops and sometimes stock. However, even while counting the cost, there's quiet comfort in the knowledge that rain means resurrection, whereas dryness and drought mean nothing but death. Droughts kill everything. Not just the feed for animals and humans, but the prospects of all the creatures, known and unknown, that share our farm. The entire property falls silent. Not only do the birds stop singing but, as the drought deepens, the flies and mosquitoes and grasshoppers decline, and you begin to miss the annoyance of insects.

You start to hear the niggling sound of worrying – the long, deep sigh of hopelessness. Yet these are emotions farmers can't afford. We must go on believing in the myth of our indomitable courage and resilience. After all, we keep reading about it in the newspapers. Bad weather can't be permitted to crush us.

The rain charts Phillip has kept assiduously ever since we arrived record that the past six months are the driest we've had since 1987. And the district's old-timers have to go back to the 1920s or '30s to find a parallel. It is now spring. June was mild, but July bitterly cold. When the cows drop their calves, there's no grass for them to turn into milk.

We entered the year with cautious confidence. Most cows were pregnant, our herd a cause for considerable pride. Extra heifers have been mated, but I now realise these young creatures won't be able to rear their calves, even though those calves are small as puppies, and as playful. Talk about blissful ignorance. As I check the animals each day, the calves dance around their weakened mothers, knocking at the udders, demanding quantities of milk that their mums have no hope of providing.

When I feel melancholy, the colour of the hills seems dull and brown. If my spirits are up, they look golden. Either way, I see a range of colours I haven't seen for years. Even the green of the angophoras and eucalypts looks unfamiliar – as if the trees judge it inappropriate to display their normal splendour. Only the acacias, notoriously optimistic, offer respite.

'It's still beautiful,' people say. But it doesn't feel beautiful when, each day, you see your cows getting thinner, your calves looking stunted, your stock bogged alongside the kangaroos in the quagmires that once were dams. Phillip and Gavin spend hours trying to extricate these creatures, only to watch in despair as they turn and sink themselves again. So we buy more solar energisers and put fences around more and more drying dams.

I'd met Gavin a few years earlier when he was managing his in-laws' organic farm near the Blue Mountains. Born on a farm in Suffolk, not far from the North Sea, his first love was and is the ocean and, prior to his marriage, he'd built yachts and captained charters exploring much of the world. He was quiet and self-reliant. Naval life had given him a mastery of machines, particularly diesel engines. But the marriage had broken up. So, needing a new job and a home while he worked on a boat that would, in a year or so, take him back to the ocean, he'd come to Elmswood to help us with repairs to our run-down gear. There's everything from tractors to chainsaws needing his attention. Phillip admires Gavin's mechanical skills; I, his interest in biodynamics. Gavin quickly settles in. But it's a hard time to be working at Elmswood, and a grim one.

Every day we check as many dams as we can. Are there dead or dying animals in them? Are the electric fences warding cattle off?

Is this dam or that dam still comparatively safe? Even spring-fed dams that had never before failed prove worthless, incapable of sustaining more than a couple of turtles.

As the dam levels go down, the diesel bills go up since I now drive 60, 70 or 80 kilometres a day around the farm, just checking things out, trying to 'triage' the crisis and ascertain which cattle are in the worst condition.

We move cattle closer to the homestead, where they can rely on troughs for water. But, in the end, we decide it is best to put them on trucks and sell them. Not all, but most. Is it better to sell a skinny cow and get less money, or a fat one and get a little more, fearful that, if we don't sell it now, it will become a thin cow in a few weeks' time? I swear there comes a time when I know every cow by heart – can recognise their faces. I stare at them and make life-or-death decisions.

'We can't sell everything,' I say. 'We can't stop breeding cattle. I'm not ready to stop.' The arguments for and against the farm spin in my head but don't lead to a conclusion. I can't give up – not after all these years. Even though there's a part of me wishing I could simply walk away and move on. Get over it.

My kitchen-table meetings with Gavin tend to begin with 'I haven't decided yet but . . . ' Or 'I've been thinking . . . ' Every decision is an agonising one. And every decision seems doomed to be the wrong one. Sell this mob? That mob? All of them? The calves? The pregnant cows? The pregnant heifers? Join some this year? Join none this year? Keep all the bulls? Sell all the bulls? Three weeks earlier, Gavin had been keen to promote feeding.

Now – after dragging cows out of dams, watching calves bashing against their mothers' udders, all but knocking them over – he is keen to see the decker parked at the loading ramp. Keen to get whatever we can off the farm.

On two occasions I make sure I'm away from the farm the day the trucks come, so I don't have to deal with it. But on returning home, I feel a sense of loss and a new wave of sadness. All those years of breeding up the herd, of making decisions on husbandry, have come to this. And the cattle we push onto trucks are probably the lucky ones. Others are too weak to make it up the ramps.

The responsibility of looking after so many animals is simply too much. In the beginning I felt a sort of determined pride that, this being our third drought, I'd be prepared. But I'm quickly reminded that every drought is different. And whilst all droughts are cruel, this one is especially so.

After selling cattle for eight consecutive weeks we still have a small mob of older cows with calves that have to go. Gavin has been feeding them hay, so they all answer to the call, not merely obedient but grateful. This makes them easier to muster but, when it comes time to bring them to the loading ramp, one cow – as if suspecting our intentions – refuses to budge. And when she does, it is to escape. Somehow she manages to get away from us and crash into the garden, where she makes a beeline for my rose bushes. Fortunately, she keeps moving and finishes up in the driveway where, despite my arm-waving and shouting, she decides to stay.

Later in the afternoon Gavin reports that the cow is still

avoiding the muster and he decides to fire up the tractor. Perhaps the threat of an oncoming machine might get her moving in the right direction. I decide to approach the beast from another angle on foot. She stands her ground before the big machine, then turns on me. I run behind a tree.

The cow, a black-baldy, twelve years old and mother of nine calves, is furious. I can see her anger, her defiance, and feel her determination to be reunited with her calf. At her age she's decided that enough is enough. She will rage, rage against the dying of the light. Despite her age and her thinness, she has enough energy to stand up to the two of us, plus the tractor.

I wave at Gavin to stop, to give up, and head home feeling shaken, barely able to walk. At the homestead I turn and look back to see that the cow has moved across the paddock towards her calf. They're now side by side, but separated by a fence. Both animals bellow unhappily. For a time she loves her calf with her head, massaging it all over, until both of them fall quiet. Then, changing position, she helps her calf nuzzle her udder, through the wire, giving it every drop she has.

Never before have I been so impressed with the power of a cow's courage and personality. No matter what we wanted, her life's work was caring for her calf, and she wanted us to know it. I can only respect and admire her as one mother to another, so I don't care if she eats *all* the shrubs in the garden. Her life is worth more than any plant I want to keep safe for aesthetic reasons. Let her have the lot, my few remaining flowers, any remaining blade of grass on the lawn.

The very nature of cattle breeding means you must harden your heart. The sunset is spectacular, but I'm in no mood to admire it. The only beauty I can appreciate, at this moment, is that of an old cow. I take a glass of wine around to the north veranda, nestle into a wicker chair and make a fuss of the dogs at my feet.

It's been only a couple of years since the foot-and-mouth disaster struck Britain, which led to millions of animals being killed. I was working in the olive grove most days when the horror stories were being reported in newspapers, on the radio and TV. Karen, a German friend who'd owned the local nursery that provided many of the plants in our garden, had been helping with olive pruning. As we worked together she told me of a time in Germany, just after World War II, when she'd gone to visit relatives farming in the foothills of the Alps. Soon after she arrived there was an outbreak of foot-and-mouth and the farm and surrounding villages were quarantined.

'There was a lot of caring going on,' Karen said, 'every farmer set up an animal hospital. Each day we would bathe the feet of the animals, nurse the kids and calves and lambs and piglets, handfeed them, ensure their comfort.'

How different things were at that time. Now bureaucrats arrive to announce the timing and scale of the slaughter. Karen couldn't begin to understand the need for all that killing, for the mass graves and the burnings, all due to a disease that doesn't spread to humans.

I hope it is a crisis I'll never have to face. Even the slow-motion disaster of the drought was getting too much for me.

I end the day with a walk to the river, accompanied by Aurora and a tumble of dogs. The poor old Pages is as dry as a cow's udder. Yet I know the river has a bubbling, surging, fluid future. Yes, it's a slight river, but I feel it has fathomless depths. I know its cadences, the variations of flow from translucence to mushroom soup. Even now, when it looks so sad, I love it. Perhaps especially now.

10

It had been silly of me to ask David Foster if he'd found coal. Around here coal finds you. Coal outcrops are visible on the banks of the Pages and both geologists and locals have known about it since white settlement; and long, long before then the Aborigines had, no doubt, found it useful.

During the Depression some of the Pages River coal had been privately mined. Manny Bloomfield, who's lived in Blandford for eight decades, remembers riding his horse past a mine in the early 1930s, when the district was infested with rabbits.

'The mountains would move in front of you. There were millions of them,' says Manny. 'We'd dig up thistle, chop up the root, put it into old Mintie tins, pour on strychnine and mix it up with our hands. Then we'd lift a sod of dirt with a mattock and put some bait down. Strychnine was sold at every corner store. It was up on the shelf next to the lollies. The rabbits ate the bait and would be dead all over the place, rotting and stinking. Thousands of crows would come and peck at them, but I'll tell you something very strange, the crows never died. You'd think they would with all that poison.'

As he crossed the Pages River with his tin of strychnine,

Manny would pass a small team of men working an underground mine at Bickham.

'There weren't many of them. I don't even think they camped there. They used a roan draughthorse to pull the skip up out of the shaft. The coal was then dumped into a lorry and taken to the railway line.'

The remains of an impressive vertical boiler that once provided steam for the mine's transport system and a haulage engine (wrought from steel by J.W. Wilson & Co., Liverpool, England) stand proudly on the edge of the Pages to this day. They guard three old shafts just metres away. You can bet there are the skeletons of quite a few animals deep below. These relics confirm that this mine was a small one, known in the industry as 'twelve-light pits', because without a fully certified mining manager no more than twelve men could be employed underground.

Coal isn't one of the earth's 3000 minerals. It's a rock; and rocks are made up of lots of minerals. However, as rocks go, it's particularly interesting and, just as you find insects trapped in amber, you can find history trapped in coal, which is why it's been described as 'nature's book.' A chunk of coal contains fossilised wood, roots, leaves, since it is formed from decomposed and compressed plant matter. It is a record of one tree on top of another, one forest on top of another.

Years ago in Ireland, Phillip and I walked over peat bogs and sank to our knees in luminous green moss. This is unformed coal, soft and damp. That's what it had been like in the Upper Hunter long ago. The whole region would have resembled Ireland's bogs.

The landscape around us isn't linear. There are no neat, well-ordered geological layers as colour-coded as trifle under our feet. Around here the earth's crust has been highly mobile, upthrusted and rotated so many times that what's old can surface whilst what's new can be deeply buried. So everywhere the older Carboniferous is lying on top of the younger Permian. It can all be blamed on the Hunter's Thrust Fault that, over millions of years, piled up the hillsides, making the coal seams dip this way and that.

It's 1791; William Bryant discovers a considerable amount of coal around Lake Macquarie. Previously a Cornish fisherman, Bryant was an escaped convict on the run with his wife, Mary. They used the coal they found to keep them warm at night. However colonial historians, being class conscious, tend to give the credit for finding coal to a Lieutenant John Shortland. Either way, within a couple of years 45 tons of coal were exported from Newcastle to India – and we've been exporting it ever since. Coal remains Australia's biggest export earner. Turn the fragile pages of 19th-century Hunter Valley newspapers and you see story after story about lock-outs, strikes, pit accidents, closures, openings and community disagreements stemming from dairymen unhappy about mining. Little has changed over 200 years, except that the arguments now require even larger teams of lawyers.

You're reminded, again and again, that the Hunter Valley's

image as a wine region may be little more than a mirage – given the dominance of coal ever since white settlement.

At the Universal Exposition of Paris in 1878, Australia mounted an impressive coal display intended to alert the world to the quality of our product. New South Wales already had thirty-three collieries and that year mined 1 575 497 tons. Mining was so extensive and growing so rapidly that the great botanist Baron Von Mueller, whose credits include the planning of the ethereally beautiful Melbourne Royal Botanic Gardens, suggested that Australia waste no time recording her bounty of flora and fauna, particularly the eucalypts, before the need and greed for coal destroyed them.

Every experienced miner in the 1800s knew the quality of coal from the different pits. And all were excited when, in 1882, a young geologist named T.W. Edgeworth David (1858–1934), who had just been appointed the geologist for the New South Wales Department of Mines, measured the biggest, richest seam imaginable at Greta. The seam was extensive, deep and jet black – and ran all the way up the Hunter Valley, from Maitland to Murrurundi. At times the Greta seam has dominated the coal market.

Edgeworth David would travel with Shackleton to the South Pole in 1907, receive a knighthood and become Australia's most famous geologist. He lived to see his work spawn scores of company towns across one of the world's great coalmining districts. He later became Professor of Geology at Sydney University, where they named the building that houses Geology after him. This building is now marked for demolition to make way for a new Law Faculty, so Edgeworth David's namesake will end up in the rubble.

Back in 1927, underground pits – some employing as many as 1200 men – were providing up to 3400 tons in an eight-hour shift. In that year, fourteen-year-old Jim Comerford went down his first pit, beginning a long career as a miner and union official. Phillip and I met Jim when he was 84 years old – at a political meeting in Kurri Kurri. Like Edgeworth David, Jim sees coal as precious.

'If carefully managed with appropriate measures it's possible to recover 90 per cent,' he said, expressing indignation at the way coal has been wasted. 'All authorities – royal commissions, mining engineers, the Miners' Federation, government geologists, including Edgeworth David – have condemned the wasteful methods used by the Big Coal shipping and steel interests.' The old miner seems to care for coal more than the mining companies that make the big profits. He was clearly enraged when he told us that, 'It could be estimated that up to 50 per cent of coal in some mines has been lost.'

Now a respected historian of the industry, Jim remembers when the quality of coal in the Upper Hunter was considered 'so low that no one wanted to mine it'. He described it as being 'too dirty, needing a lot of processing before it was any use'. The dirty coal is put through a washery and cleaned before it's loaded on to trains for export. It wouldn't need to be washed if it wasn't destined for furnaces to generate electricity. It could be recovered (that's the official word for extracting) for use in the manufacture of cement, bricks or tiles, for which Jim thinks it is far better suited.

The last time we saw Jim he was about to turn 90, and was as irascible and inspiring as ever. He reminded us that Australia's history with coal began with James Cook, who wanted a common collier with a shallow draft and wide body to chart the uncharted oceans. Moreover, when Cook got into strife on the Barrier Reef and had to jettison a lot that the *Endeavour* was carrying, including some of its cannons, he refused to part with his good English coal.

Jim Cook and Jim Comerford would have understood each other.

'Libraries aren't just for books you've read, Patrice,' Phillip says. Which explains why he has tens of thousands of them in his library: not only every book he's ever read, going back to William books, but every book he intends to read. When we moved into Elmswood, first one, then two bedrooms in the house were immediately filled with bookshelves – on all walls, floor to ceiling – and it continued to grow until every other room was half-filled. My books, only a fraction in number, went upstairs into a spare bedroom, now Aurora's. Unpacking them, we discovered that we'd bought the same editions of many books, but neither of us was willing to part with ours. Determined to keep my books separate from Phillip's (for fear of never finding them again), I put floor-to-ceiling shelves in my office. This is just as well, given that Phillip proceeded to colonise the new barn.

Visitors, daunted by all the antiquities, say, 'You must feel like you're living in a museum.' It feels more like we're living in a vast second-hand bookshop.

The various libraries now have separate names: the Main Library downstairs; the Antiquity Library upstairs; the Barn Library; the Dining Room and Living Room libraries; Aurora's Library; and, fiercely protected from Phillip's imperialism, My Library.

I often wander through his, just looking at the bindings, perhaps reading the inscription written by an author. Or I'll open an original edition of Voltaire's *Candide* or Darwin's *The Descent of Man*. And because Phillip jumbles books together, I'm forever shuffling them around, so that Camus is closer to Sartre and not next to P.G. Wodehouse – whom Aurora has recently discovered, taking a liking to the character Jeeves. Every time I do this, it makes Phillip cross, because he somehow knows where most of his books are, or should be. So I retreat to My Library and move my books around.

There's more and more talk about banning grazing on river banks. I know the theories: the cattle trample the edges and pollute with their waste. Yet at Elmswood, we've never been able to bring ourselves to fence *all* the river frontage off from stock. And that's not to save us work and expense, because allowing cattle access to the river means we have to build flood crossings to keep them in the right paddocks; and flood crossings are wrecked as soon as the first rush of water moves downstream.

We try not to overgraze the one area where cattle are allowed. Much of our section of the river hasn't had cattle near it for almost a decade. However, the cattle do make one useful contribution; they help control weeds. Since we refuse to spray herbicides, we use cattle to graze the young shoots of Noogoora burr, which proliferates after flooding in summer. Noogoora burr grows like Jack's beanstalk and, whilst not quite as venomous as Bathurst burr, can become a major problem. Sometimes its germination is so quick that we aren't able to manipulate the herd in the right numbers in time to do the job properly. It's a weed that needs to be grazed hard and fast. We also use cattle to prevent grass from becoming a fire hazard and a snake hazard around the pumps.

Weeds are seen as a benchmark of a farmer's care for the land. A good landowner has no weeds. Consequently most people think nothing of hitting the weeds with massive amounts of chemicals – sometimes the defoliation makes me think of Agent Orange in Vietnam. And a lot of the spraying takes place along the edges of our rivers where it's inevitable chemicals will leach into the water.

Check the legislation and you'll find that most chemicals cannot legally be sprayed along watercourses. This doesn't stop weed inspectors recommending, even demanding, spraying. And graziers oblige. How did the weeds get there in the first place?

Usually bad management practices or as a result of garden plants going feral.

There are, however, a few facts about weeds everyone agrees on. First, they're everywhere. Secondly, despite decades of spraying, old weeds are staying and developing resistance to herbicides, while new ones are always arriving. Thirdly, it's agreed that weeds are expensive, costing the nation at least $4 billion a year in lost production and chemical warfare.

Increasingly landowners and scientists are coming to the conclusion that weeds of one sort or another are here to stay – and we're better off aiming to lower their population levels over the long term than trying the annual Vietnam-War approach.

In the suburban homes of Australia, the chemical blitz against every sort of bug and microbe, real or imagined, is possibly leading to a less-resistant human population. It turns out to be healthier to live with a reasonable level of germs. That's my attitude to weeds. We graze them where we can. We chip them, hoeing away in the sun, when we have to. But eliminating them is utterly impossible.

Yes, they can break your heart – and cause other coronary problems. The local real-estate agent tells me what drives elderly farmers off their land isn't so much drought or poor cattle prices as the war against weeds.

Our latest aggravation at Elmswood is tiger pear. 'Tenacious buggers,' says Phillip, trying to extricate one that has penetrated the thick sole of his boot. Others have punctured the tyres of his

ATV (All Terrain Vehicle). Once again, the floods introduce them and they climb up the river banks and hide amongst the grasses. Cattle spread them across a paddock, whilst kangaroos take them even further afield. You find them piled up along the fences where the roos wriggle through. Given a large clump, cochineal bugs will have a munch at them – but most of the time there's just a little bit here and little bit there. Thick leather gloves give no protection, so when Phillip hoes them, he picks them up with the tongs from the barbecue and, when he's got a truckful, carefully, delicately places them on bonfires, rejoicing as they burn.

While driving around the Pages in December with Michael, who lives upstream, we bounce over a recently ploughed paddock, aghast at the Bathurst burr germinating like a crop. It's a weed that loves colonising disturbed soils. After we cleaned our dams, the walls were covered in them. Michael will kill most of these burrs simply by reploughing the paddock before sowing oats. He tells me of a neighbour, further into the hills, who has been spraying them for forty years – an immense investment in time, labour and chemicals. Yet every year exactly the same number appear, so Michael has finally come to the conclusion that it's easier to put up with them.

When I am most troubled by an infestation of this weed or that burr, almost all of which you first notice along the river, I ring my friend Stuart, who, after listening sympathetically, invariably says something wise.

'The rivers, like the railway lines and the roads, are all "commons",' he'll say. 'They carry ideas as well as unwanted seeds.

No individual can be expected to heal the damage of their common, because the problems don't begin and end on their land. The common is a carrier of good and bad.'

Two kilometres downstream from the Capuchin's hermitage a pair of upright cylinders, three metres in diameter, are positioned at the edge of the Pages. They're used to redirect water, by gravity, into a 10-inch underground pipe that leads to a holding dam 10 kilometres downstream. The system was put in place in 1978 to provide water security for the township of Murrurundi, which had, in the past, frequently gone thirsty.

The dam holds 160 megalitres and has an impressive wall covered in grass that's mowed regularly – not just for appearances, but to ensure there are no leaks. In a crisis it can provide a managed supply of clean water for nine dry months. Yet it's nothing compared to some of the dams over the range that feed the crops on the Liverpool Plains. There a dam isn't a dam unless it can hold at least 500 megalitres syphoned out of flooding gullies with diesel pumps the size of train engines. They make Elmswood's struggle to hold 13 megalitres seem pitiful.

It's the summer of 2002 and Murrurundi's dam – now a pond – has become infected with toxic blue-green algae called *Aphanizomenon*. In this emergency, water is being taken directly from what's left of the river.

'It was like the water had smoke through it,' says Phillip Dunn, the Murrurundi Shire Council's engineer. 'We thought the water had "turned".'

A place called the 'blue hole' has been dug out mid-stream in the Pages at Murrurundi, so that water can be pumped into holding tanks above town. Despite a wire fence around the hole and a Keep Out sign, it's popular with swimmers. An excavator took three days to dig it and now a 25-year-old Hanson Sykes Jetting Pump – 'a magnificent piece of equipment,' says Dunn, patting it – hammers away alongside, taking 300 000 litres a day to distribute to a thousand locals.

Aphanizomenon belongs to the Nostocales order of algae. They're one of the clever plants you can't help but admire. They don't rely on any external nitrogen source, simply extracting from the air (like a legume). They don't look toxic, unlike the algae that terrified the nation when it choked the Darling a few years back, burning it bright blue. Although it's a close relative, it isn't killing the ducks and fish in the Murrurundi dam, yet it represents a danger in drinking water.

So how did it get there? The algae likes phosphorus, and the land around Murrurundi is a Tertiary basalt, rich in apatite, which, in turn, is rich in phosphorus. If the apatite soils are responsible for the high phosphorus levels in the water, why wasn't it a problem before? The Pages has always flowed over the same land. Yes, flowing is what it's been doing for at least forty million years after the rocks began to lift.

It emerges that today's problem began in 1978, when dam

storage was built. For what algae likes most is still water. When the water is flowing, the algae just keeps on moving, never getting a chance to build up, but when it ceases to live that nomadic life (such as in a dam) the algae settles down and starts blooming. The aerial spreading of superphosphates across the hills near the headwaters hasn't helped either.

Usually pumps extract dam water 600 millimetres below the surface, where it's believed the best water is. After the thousand residents of Murrurundi are done with it, it ends up swishing around sewage ponds for sixteen days before re-entering the Pages and continuing along its merry way.

'Has the algae spread?' I ask Dunn as we stand above the dam watching the ducks. Apparently not.

'No, the damn stuff is only in the dam.'

There's no easy way to treat it. The shire could throw a lot of copper sulphate into the water. That would quickly kill it, but it would also kill the fish.

There's no state-of-the-art filtration system available either. Even if there was, how could a small rural shire possibly afford it? Murrurundi gets just a third of its funding from rates and two-thirds from government grants – a total of $5 million a year to look after the infrastructure spread over 2471 square kilometres, including 550 kilometres of road.

So the council's decided to wait, living in the hope that the algae will simply go away. On the advice of the DLWC, it should die of its own accord, taking approximately six weeks. But when I first visit the dam it's been blooming for thirteen weeks.

Meantime, as anxiety mounts, the council is fielding calls for water to fill up domestic tanks.

It had better rain.

11

It was seeing it in writing, in the local paper, that finally forced people to confront the issue. The coal company had organised a private meeting with the Murrurundi Council. Ominously, it was closed to the public – and open to the local media only after widespread protests.

This led to an anomalous situation. The media was in; the public was out. Yet the *Local Government Act* clearly states if a council is being briefed and the media is present, then the public have the right to be there as well.

Did the council stipulate this out of ignorance? Either way, it wasn't a good start: an illegal meeting, chaired by a mayor already loudly endorsing the project, despite knowing little about it. At first the company was promising as many as sixty local jobs, but when subsequently interrogated, conceded that it could be as few as three.

And things didn't improve during the next few months.

The coal company gave a tour of the site to a select group of locals, claiming to be 'fair dinkum' about listening to 'people's concerns.' Dismissing the paddock chosen for the mine as 'goat country' and 'good for nothing', they clearly revealed their ignorance of the district, their land and part of the river.

It took eight months for any details to be put on the table – in the form of a 'Review of Environmental Factors', a fat and fatuous document that set out the mine's plan to take a 25 000-tonne bulk sample.

I asked geologist Emeritus Professor John Roberts to help make sense of the report. Though it was prepared by a team of alleged experts, it was abysmal. In his opinion, 'If this was submitted by one of my students, I'd fail them.'

There were discrepancies between the test sites and their positions on the map. Vital geological information was missing, yet the report had room for some preposterous claims, such as the insistence that the river wasn't connected to the adjacent aquifer. Their salinity readings in the river were high when every other survey ever undertaken had stated the exact opposite.

So the emeritus professor sent a fax. To extract a 25 000-tonne coal sample that goes deeper than the riverbed and is a mere 100 metres from the river could cause the river to divert into that hole – and eventually, into the mine site. In other words, it could 'capture' the river. John asks hydrologists to confirm his assumptions. They do.

I find 100-year-old maps showing the coal seam crossing the Pages and heading north up over Timor Road. It's identified as part of the great Greta seam, and is called the Koogah Formation, laid down in the Permian period about 250 million years ago. Further searches in the archives turn up surveys from the 1940s describing the coal as 'good' and suggesting open-cut methods would be the only way to extract it. However, it is noted, there will be problems: 'The fact that the Pages River flows over the outcrop would probably mean that considerable quantities of water would be encountered in any workings.' And that, surely, is the principal worry at the beginning of the 21st century.

Perhaps a two-hectare hole in a hill isn't much to worry about. But it's not the width of the hole that's the problem. It's the depth. It will need to go down to the lowest and thickest coal seam, called the 'G-seam', 20 metres below the riverbed and straight through the aquifer. This means the water (which the rest of the community needs) will represent a problem to the engineers, who will most likely regard it as a waste product. They'll have to get the water out, so they can extract the coal. When I hear they propose pumping up to 250 megalitres and spraying it on the road 'for dust suppression', I'm appalled. Those of us on water restrictions for months see water used for dust suppression in a drought as nothing short of sacrilege. Moreover, they're talking about a volume of water that is half as much again as is allocated to the entire township of Murrurundi.

How could this happen when there's a moratorium on water licences?

There's a rumble of disapproval through the community and, just four weeks later, Bickham Coal change their mind and say, 'We'll pump the water out of the aquifer and drill holes 500 metres south and put the water straight back in.' Somehow that makes their attitude to water seem even more casual, more cavalier.

So our river will divert into a coalmine? Disappear? If this were the case, then clearly the proposal shouldn't go ahead. This is the deal-breaker, the showstopper. The Department of Mineral Resources would have to use its veto and what minister would like to go down in history as the politician who 'disappeared' an ancient and lovely river?

Silly me, I forgot. It's a coalmine. It's export dollars.

Back in June 2002, before I'd developed my angry, anti-Bickham coalmine position, I'd had a fantasy inspired by Swedish doctor Karl-Henrik Robèrt, a most remarkable man articulating a vision of natural resource management. Having followed his work for years, I drove to Sydney, to UNSW to hear him outline 'The Natural Step', an approach to nature that strives to build consensus.

The essence of The Natural Step turns traditional debate upside down and asks: What do we agree on? How can that knowledge serve as a more solid platform for community building? Many a company has used the system to assess if its business is moving towards a more sustainable future; and many have modified or transformed their activities as a consequence.

Perhaps here was a chance for us, as a community, to assess the coalmine proposal using The Natural Step's framework. I'd been talking to one of Karl-Henrik Robèrt's colleagues, Dr Joe Herbertson, a Hunter Valley local who has spent his entire career working in minerals and metallurgy, including thirteen years as a BHP research scientist.

Joe was enthusiastic at the prospect. I wanted proof that The Natural Step was more than lofty idealism and he wanted consensus-building applied *before* a project was up and running. We both detested the warfare approach to development whereby a company defiantly announces a project, while environmentalists invariably oppose, going nose-to-nose, eyeball-to-eyeball, hammer and tongs, with the winner being the last one standing.

During one frantic day, Joe and I meet separately with local anti-mine residents, the council and David Foster from Bickham Coal to brief them on the concept. For it to work, we'd need everyone to willingly participate. We thought, naively, that Bickham Coal could use this procedure as part of its commitment to a 'community consultative process' and as evidence that they were, as they claimed, 'fair dinkum.'

No one actually says no to the idea and some of the councillors seem enthusiastic. After all, The Natural Step could provide a development framework for the future – in a shire that until recently required no development applications whatsoever. Moreover, they've endorsed a vision splendid for the shire under the banner of Murra County – a place for clean, green living.

It becomes clear Bickham Coal isn't interested, however. Unanswered phone calls are a giveaway; ignored correspondence outright rude. Instead the then Minister for Mineral Resources, Mr Eddie Obeid, responds by appointing his own consultative committee to address community concerns. After all, a state election is in the offing.

Soon afterwards the fraud of official 'consultation' reaches new heights of farce. I thought consultation meant 'to consult', and that the mining company was directed by the Minister to consult with us, the community. But judging by the meetings held so far, this isn't the case. All the major issues that the community has raised for discussion remain unresolved.

Hundreds of people over a period of months have written letters to the Department of Mineral Resources expressing concern about the mine, yet the department refuses to reveal the number of submissions, let alone their contents. After five months of silence and little progress we've gone back to the bureaucracy again, requesting that the purpose and power of the consultative committee be clearly stated in writing. We receive nothing. No one in head office knows what the committees are for. Nothing is in writing. No one can state with certainty why they exist.

Every mine in the Hunter Valley has such a committee and everywhere people describe them as a waste of time. They may have been founded with the best of intentions, but they are now seen as just another obstacle to overcome. So when the department sends out press releases proudly proclaiming 'consultation', those who've been trying unsuccessfully to consult are bound to regard it as an insult.

I ask David Foster if I can walk along the Pages River near the proposed site. Starting upstream from the Wrights' graves, the river soon takes a 90-degree turn and cuts through a magnificent craggy gorge, even narrower than Cameron's, which forces me up and out. This is the most beautiful part of the Pages. Climbing high, I'm surrounded by a thousand shades of brown and green, so dry that you fear spontaneous combustion. Even the friction of my shirt against a branch could start a fire. The only sound is the crush of dry leaves underfoot.

Like Samson pushing at the columns of the temple, trees

shoulder apart boulders and overshadow deep aqua rock pools that are the last repository of the diminished river. For a moment, the drought feels distant. If water could think, it would be warning itself to stay here, safe in the pools, for to journey onwards will certainly lead to death by evaporation. It's hard to believe that this is the same stream that's long since stopped flowing at Elmswood. For 400 metres the gorge directs the river eastward before it ushers it southward, meandering past a hill that will largely disappear if the mine begins bulk extraction.

This development is madness. It has to be the craziest idea in the world.

The 920 residents of Murrurundi township have been on water restrictions for months. The algae hasn't died off and the famously reliable Blue Hole is getting smaller by the day.

Finally, desperately, the council decides to wage war on the algae with the ominously named Cupricide, and within ten days the water is clear and passes Health Department tests. Dam water can once again start flowing through the township's taps. Ironically, water restrictions and algae have ensured a full dam is available, but maintaining an adequate supply of town water isn't off the agenda yet. And you can hear a murmuring countdown. A hundred days of water left, 99, 98 . . .

The year 2000, the Olympic Year, was very wet, and the river flowed clearly and abundantly for months.

Although we'd begun the process of water reform, we knew the plan would be designed on the worst-flow levels – and they were only the worst in our collective memories. God knows how bad they'd been in the distant Aboriginal past.

For the time being it wasn't a problem – the problem was Phillip. He'd been unwell for months, really sick for the first time in our years together. Oh, both of us had banged knees, bruised elbows, taken falls and felt the deep fatigue of too much work. Now it was something serious and, after endless inconclusive and contradictory tests, the specialist told him that his heart was faulty – and I had the phone call preparing me for the worst-case scenario.

I couldn't believe my big-hearted man had a dicky heart. It seemed impossible. Just when they were planning to saw him open for a multiple bypass, it was discovered that the real culprit was his thyroid. What followed was harrowing and dramatic. Things went wrong on the operating table and in intensive care, but with Phillip's spirit and my yelling at the doctors, we pulled through.

Back here at the farm it's wet, as is the whole of eastern Australia. We're packing our suitcases for a busy three weeks. Despite his wobbly health, Phillip is heading for Perth and Brisbane to give speeches and then, and only then, will he surrender to the surgeons.

Reports of flooding across New South Wales and Victoria are

big news, and our little river is rising too. By Saturday afternoon, Phillip has parked his car across the bridge, on the other side, to prevent being flooded in. Just before dark he decides it's time to head back to Sydney – he's got enough problems without getting marooned. We agree to meet in the city when the flood is over. Aurora and I, accompanied by two dogs, wave goodbye at the river that's beginning to roar. We can't hear each other even when we shout. Ducks float past at record-breaking speed, seeming to thoroughly enjoy themselves. Do they care where they'll end up?

We trudge downstream and people travelling on the road above us wave out their windows. Before dark we return to the homestead for torches and resume our explorations. The dogs run barking ahead of us, frogs are hopping everywhere and a few excited turtles speed by.

Returning to the bridge, it's clear Phillip left just in time. Aurora insists we turn off the torches. She longs to be cut off from the world, to be an island for a day or two. Because, of course, it will mean she can miss school. Now nothing can be seen. Clouds hide the moon. Only the river can be heard and felt as its floodwaters thunder by.

I'm afraid of this deafening darkness. Surely this is what drowning must feel like. I turn my torch back on and, yes, the bridge has drowned. A big tree, swept downstream, is wedged against it, sending sizeable waves towards us. We retreat to higher ground, yet still I don't feel safe. Normally I'd put a marker at the water's edge, to see how fast the river is rising. But it seems too dangerous – as though the Pages might give a mighty surge and sweep us away . . . I scream when I see Aurora nonchalantly walking closer.

Sweeping our torch beams around, we look for the last of the trees we planted on the banks. We look to the left, then a little to the right. Yes, they're still above the flow, but about to disappear. And, yes, the giant irrigation pump has been dragged up enough. Finally, exhausted by the flood's thunder, drenched by the rain, we head for home. It's time for a hot drink and bed.

An hour later, I snuggle under the blankets and try to persuade myself to sleep – but I can't stop worrying about the river. It could never reach the homestead or the out-buildings, but there's still danger in its sound. I remember past floods, when the river rose even higher; when I heard the noise of its depredations as it wrenched she-oak after she-oak from the river's edge below the olive grove. We'd heard a crashing and, in disbelief, ran across in time to see trees falling like pins in a bowling alley. One after another they had tumbled.

I remember feeling sorry for the ducks that had lived amongst them. Worrying how we could ever establish trees that close to the river's edge again. Wouldn't every successive flood just rip them out?

I recall standing near the edge, watching the water rise to an unprecedented height, to where the giant bronze statue of Mercury now stands at the second cattle grid. Tonight's flood is at least 8 metres below that level. So what am I worrying about? This is just an average, ordinary flood. But I can't slip into sleep. What will I discover in the morning?

I wake and realise that the flood hasn't abated – that the mighty sound of the Isis and the Pages colliding below is, if anything, louder, but I won't be able to see what's happening until there's a hint of sunlight. So, just before dawn, I stand shivering on the veranda and, looking across the lucerne, see that both rivers are rampaging with equal force. This is a rage that isn't going to let up, and the clouds are still bloated.

I decide to let Aurora sleep in, to make tea and return upstairs to get a better view. But no sooner have I sat down in the kitchen, waiting for the kettle to boil, than Aurora is beside me. Fully dressed. Wanting to be outside. I can't let her go alone, though normally I would, with a stick in her hand and the dogs for company.

Again we rug up, put on gumboots and slop down the same path as last night. Yes, the flood has begun to retreat. Some of the bridge is revealed but, as usual in a flood, both ends have been washed away and it is impassable. The silt dumped on our side of the bridge must be 2 metres high. The smoothness invites a footprint. So over we go, leaving our mark: 'We were here first.'

Where have all the birds gone? What happened to the ducks? Then the bittern sings out and Aurora, walking off on her own, sings with it.

Now, strolling beside the river – full of topsoil, branches, fence posts and miscellaneous muck – I remember something I should have remembered last night. I'd forgotten one thing, something I often forget: to turn off the pipe that carries water to a trough across the river in a paddock called Pennyroyal. If the flood gets high enough, it pulls it down and water from the tank simply pours out. And if that's happened this time, it'll mean there's no water in the tank, no water for the toilets or the garden taps until the flood is entirely over and until the river is hushed, calm and unmuddied – because you can't turn the pumps on until the flow is clear.

Most important of all, there'll be no water in the troughs for the cattle. I shout for Aurora to come quickly – and we run back to the homestead for the Gator, a vehicle with six fat wheels and a tip tray that serves as a giant wheelbarrow. And, thanks to its big, spongy tyres, it can get over the upheaval of rough rocks and slip and slide through the thickest mud. 'Can I drive, Mum?' asks Aurora. 'Sure.' And we head towards the forgotten pipe, much of which has been lost to the flood.

Then it's uphill to check the water tanks which, yes, are almost empty. But not quite. There's just enough for the cattle for at least a couple of days until, hopefully, we can shove and manoeuvre the pumps back in.

We breakfast, change into dry clothes and then go outside to get wet again. We can feel the sun trying to get through the clouds and, under the Drizabones, we already feel musty. It's still drizzling, but the clouds occasionally part – and when shafts of sunlight reach the ground we're convinced we can actually see the grass growing. The lucerne stands back up, as in a time-lapse sequence in a David Attenborough documentary.

Worms are writhing in every puddle. Silver snail slime coats the vegetables. Butterflies prance in between showers and flies descend on our backs whenever the drizzle eases.

Our old border-collie–Scotch-collie cross, Rosie, finds a bleached bone donated by the waters and Molly, the younger Jack Russell, tries to take it off her.

Aurora wears a big, thick jacket punctuated with six zipper pockets and, one by one, fills each of them with rocks. For today's collection, she focuses on those with crystals.

A new picture is forming where the Isis meets the Pages – and, shutting her eyes and cupping her ears, Aurora insists the sound is like the ocean. She demands I follow suit and, yes, she's right. We could be by the Pacific or the Atlantic.

This is what we love most about our little rivers and the coastline they provide to Elmswood. Even when they're not being dramatic, they're immensely entertaining – and, like the farm itself, utterly unpredictable. The farm and the rivers are like interesting but plotless novels where you're absorbed in the flow of narrative rather than drawn to a conclusion.

Before dark, I examine the slosh in the veggie garden and decide to harvest all the beetroot, and rescue the lettuces from the slugs.

The excitement of a flood gives way to a faint, flat feeling of weakness. It isn't just the roar of the waters that leaves you feeling powerless and insignificant. It's also the fact that floods leave so much work to do in their wake. Work we hadn't allowed for or scheduled. Tasks are endless, Sisyphean. Now there'll be river crossings to tackle, trying to extricate the broken wires from the silt and debris in order to reattach them to the trees that double as fence posts.

Three days later I go to back the 4WD out of the garage and discover the key is missing. I search the obvious places in the kitchen and it isn't in any of them. So I look in the secret place that only Phillip and I know about. It's not there either. Finally, I phone Phillip, and there's a long silence before he confesses that he has just found them in his pocket. He'd moved the truck into the garage, locked it and left for Sydney. Now he's in Perth, about to give a lecture to a couple of thousand people.

'Are you still wearing the same shirt?'

Long silence.

'If you are, change it before you go to Brisbane.'

Meanwhile, I'm not going anywhere. Nor, to her delight, is Aurora. Because the bridge awaits the arrival of the municipal bulldozers that will be working their way towards us, dealing with other flood damage on their way.

The rains of November 2000 stripped some olive trees of their first flowers, reducing our harvest. It will be the same for the whole of the Hunter Valley, where bruised blossoms will float along the many tributaries of the Hunter catchment, heading somewhere out to sea, past Newcastle. That rain, six solid days of it, 11 inches overflowing the gauges, made it our wettest November on record, equivalent to half the annual average rainfall.

The unexpected migration of the olive blossoms reminds me that rain and floods make a mockery of the idea genetically engineered (GE) seed can be contained within negotiated quarantine areas. Beyond the Pages, the floods soak entire valleys, shires, districts, towns. Just as they'll spread silt and weeds, they'll spread the seeds of crops, from traditional wheat to GE canola.

So while one part of the grove has enjoyed a good, deep drink, another is all but ruined. I thought I was brave to plant a test grove in black basalt, but now feel that it was complete foolishness. On the hillside, the olive trees are growing spectacularly, despite the frost of the previous winter, but the low-lying sections look like seaweed. Still, they're not quite dead.

But as I walk down a row and shake the water from a bough, I see that something else is very much alive. A million bugs lift into the sky, before settling back down again. Oh no, the dreaded lace bugs.

I've since discovered that a small tree that grows in most gullies on the farm, *Notelaea microcarpa,* commonly called 'native olive', has always played host to lace bugs that munch its leaves. Little did I realise that when we began our olive grove, we'd also be planting a 3000-tree feast for them.

Despite having 10 000 acres of land behind us, I can still start to suffocate in the homestead's domesticity. I feel the same feelings, a sort of comfortable claustrophobia, when visiting homesteads far more isolated than ours. So the longer I live at Elmswood, the more I like to spend time away from the house.

Other farmhouses are well hidden, almost secretive, but once pride takes over and people cultivate flowers, landscape a barbecue area, aim the satellite dish on the roof, plug in their pressure pump and load the dishwasher, they become much like suburban homes. Even before you play a CD or turn on the telly, it vibrates to the hum of modern life. So it's necessary to escape. Sometimes even the company of a horse is too much. The best company is the land itself.

A woman, new to the area, tells me, 'I am a people person'. Code for 'I don't like all this time by myself.' She's keeping herself busy, gardening and planting trees and handfeeding cattle, but this hasn't satisfied her need to be with people, to be involved. There are numerous service clubs – just observe the plaques worn like medals on the display boards outside most country towns. But to get to the meetings and functions can mean hours on the road.

I prefer to walk around the paddocks, along the river, and then amble home, where I rediscover that domesticity has its pleasures: playing with Aurora; watering my vegetable garden; cooking.

12

There's a surveyor's site – a geodetic station, marked on maps with a little triangle – on the summit of Black Mountain. Looking a bit like a sundial, there's a spherical design on top of a cement stand and a small New South Wales Mapping Authority plaque fixed to one side. There is no other information. It doesn't bother to tell you, for example, that you're 1022 metres high. Which is pretty high, given the flatness of our ancient continent.

The site has been fenced off, along with a mysterious demountable building that has a communication tower stuck on the top and a few radiation-warning symbols. A solar panel has been disconnected and now powerlines connect the building to the grid. An eerie hum comes from within. When we want to stand on the official top of Black Mountain at the geodetic station, we have to climb through the fence, feeling like trespassers.

Our distant boundary fences are the most difficult to manage, not simply because they're so far from home, but because they run up and down as alarmingly as graphs recording economic performance. Phillip and I like to do quick reconnoitres on the four-wheel bike, to keep tabs on their integrity or, more accurately, their frailty. Murphy's Law applies to remote fences, for if a tree or a heavy branch decides to crash, it will invariably do so over a fence, thus creating an open invitation to cattle. Even without falling trees, there's the impact of kangaroos that prefer to burrow through fences rather than jump over them. Feral pigs don't help, nor rogue bulls.

From the top, we can see the Pages on one side and, on the other, the New England Highway where it cuts through the irrigated valley of Kingdon Ponds. Clearly visible is Scone, and the racetrack with its Chinaman's hat pavilion. Much closer is Parkville, with its chook sheds and the infamous piggery, once co-owned by Paul Keating. Adding to its fame and notoriety is the image of philosopher and animal liberationist Peter Singer imitating the suffragettes by chaining himself to a railing – in this case a railing surrounding a pigpen. The curse of the piggery led to a political embarrassment for Keating and a brief time in prison for Peter Singer. These days Elenium Pty Ltd, an agricultural commodity company, owns it. (Although it's since gone into voluntary administration and is now in liquidation and seeking a financial saviour.)

The 40-hectare piggery has been part of Parkville since 1971 and employs thirty locals full time. Back in 1991 the then owners, Hatton and Brown, decided they wanted to make it the biggest piggery in Australia. It would be 'high-tech', 'state-of-the-art'; it would follow 'best practice' and produce the best pigs. In spite of the superlatives, we're happy to have Black Mountain between us and the piggery (out of sight, but not out of mind), because it's clear the water from the venture must inevitably drain into the gullies below and into the Kingdon Ponds aquifer.

On many of his geological trips to the district, John Roberts would be asked if it were true that the Pages once flowed down through Kingdon Ponds. Looking at a map, it seems exceedingly odd that the river bothered to cut through gorges to get to Gundy, when it could more easily have flowed straight down a wide valley.

'The theory is possible,' says John. 'All that had to happen for the Pages to change course was that thousands of years ago land was uplifted in one place or depressed in another. Remember that, when the course of the river probably changed, the topography could have been flatter than at present.'

When Henry Dangar first surveyed them in the 1820s, Kingdon Ponds was a cluster of waterholes north of Scone, with the whole valley as its flood plain. Over the years, all the ponds have silted up and Scone now seems safe from inundation. The flood plains no longer catch the silt that created the 'friable' earth written about by the first white settlers. Instead they're covered with farms and houses, many of which are downwind from the piggery. Residents within a 6-kilometre radius have been complaining about the stench and the run-off ever since it was built.

In the winter of 2001 the piggery, having spent $5 million upgrading the facility, invited locals to a 'show and tell'. This would be the first time many of us had set foot on the accursed premises. A 4-metre-high fence surrounds the pigpens, the effluent irrigation and the composting areas. Before we're allowed to enter, it's mandatory to don white overalls, a paper shower cap and rubber boots. We are, after all, about to enter a quarantine zone. It isn't clear whether the gear is to protect us from the pigs or vice versa.

First stop is the sows, crammed into individual pens in sheds where drawn blinds ward off all natural light. They stand on grates to allow their effluent to be flushed down into channels under the sheds, where it ends up in cement tanks before being sluiced and used to irrigate crops on the adjacent paddocks. Here the sows wait until they're about to give birth, at which time they are moved to a shed with larger pens. We walk on, past more and more sheds in this porcine concentration camp, and I say to Stuart, a vegetarian, walking beside me, 'Why waste ten years of effort to reduce the odours, when the battle should have been to close this awful place?'

For both of us, raising animals like this is monstrous. As a beef producer, I consider this kind of animal husbandry closer to making cars than farming. As the tour continues, I find myself holding back tears. In the birthing sheds there are dead piglets in most of the pens, crushed by their constrained mothers. Dead piglets are a mere statistic to the manager who tells us, without a hint of embarrassment, that 10 to 12 per cent mortality is 'average' and that all the dead piglets 'get composted in a heap further up the hill'. Dead pigs are par for the course when 2400 sows drop, on average, ten piglets twice a year.

We walk to the weaner sheds and, when the blinds are rolled up, hundreds of piglets rush to us, squealing. A sour stink pervades everything. Manure mounds 10 metres high rot down before being sold to local farmers, who use the muck as fertiliser. Once again, I am appalled and disgusted.

We walk around the effluent ponds, where massive pumps are monitored by expensive machines. They are built alongside the old infrastructure, and it's obvious why the EPA and New South Wales Agriculture consider the site inappropriate for a piggery. This place was built before environmental impact studies. Today, I believe, a development application for this piggery would be turned down.

Gullies, choked with ancient rubbish, run through and around the piggery, from shed to drainage and irrigation areas. They speak of the history of the business and show that no one gives a damn about how things look. Out here it's a veritable pigsty. And what's the justification for this horror show and all of the pollution? Cheap pork, mostly for export.

Parkville pays for the privilege of having a piggery by having its 'commons' (the surrounding air and water) polluted. Surely 'commons' should be for the *common* good, not the *company* good. All efforts to reduce the pollution of air and water have come to nothing – and it seems that no one has the power, courage or inclination to stop them. (In June 2003, it is revealed that Elenium Pty Ltd was fined a token $1500 by the EPA for 'emitting offensive odours' back in 2002.)

This is where our system of regulations becomes farcical. The regulators include the EPA, Scone Council, DLWC, and New South Wales Agriculture. Individually and collectively they fail us. Moreover, everyone knows it. Including former federal environment minister, Senator Robert Hill. He spoke of the absurdity of measuring the productivity of agriculture according to the produce per hectare when the price of degrading soil, water and air isn't included in the calculations. Every economist knows that we – Australia, the world – generate 'profits' through the loss of the natural asset base we share. Even Amartya Sen, a recent Nobel Laureate for economics and Master at Trinity College, Cambridge, believes that ordinary people aren't able to access their entitlement to basic necessities – such as the people of Parkville, who have been denied access to what is rightfully theirs.

Before plans to extend and modernise the piggery, there'd been complaints about the pong. All sorts of farms, everywhere, stink a bit. Put 200 head of cattle in the yards after a wet week and the smell is intense. When our sheep huddle under the barn on wet days their stink comes through the floorboards, and Phillip exits gasping. Even making compost stinks. The issue with the piggery is that its stink spreads so far and wide, even to the northern end of Scone's residential area.

Scone Council is charged with the responsibility of ensuring the piggery manages its waste, yet after a decade of complaints and hundreds of graphs monitoring stink levels, nothing has changed. In 2001 the EPA resumed this task, and still more forms and discussions and graphs resulted. And again nothing changed. The EPA has now granted the piggery a licence, with a long list of conditions to ensure there's no pollution and that the concentration of waste and odour is contained within their boundaries.

All this must be monitored and recorded. When local environmentalists held a meeting with the EPA to find out *how* the EPA would ensure these conditions were carried out, they were told that the locals (i.e. those who gag from the stench) would be the 'policing agents'. Scone councillor Peter Hodges, a local resident, went on ABC Radio and, with desperation in his voice, demanded the piggery be closed because clearly no one was able to stop it from smelling.

Even worse, the threat to underground water around Parkville, part of the Kingdon Ponds catchment, had been alarming local officers of the DLWC in the 1990s. It was their highlighting of nitrate contamination that first alerted me to the pollution of our local water system. Due to the hubris of the piggery's earlier owners, who were determined to grow the business tenfold, the DLWC undertook more detailed water sampling and discovered that all across the aquifer, downstream of the piggery, nitrate levels were way above acceptable limits. The piggery had been polluting the underground water supply for years, yet no one had actually known this was the case.

※

Finally we get approval to move the tree, but can't find anyone to do it. Meanwhile Col knocks on the door, wanting to know if he can use our backhoe to spread bentonite in the crater of the leaking dam. He and Terry have been shovelling the heavy white mineral by hand and they're getting nowhere.

'Of course,' I say, 'do you know how to use it?'

Col shakes his head. 'We've got licences to drive machines ten times bigger than that!' he says. He's not offended; it simply confirms his view that I know nothing about machines.

Four hours later, Col returns the backhoe and I seize the opportunity. With thinly veiled reluctance, he agrees to bring the bulldozer down and help with the she-oak. Aurora stops trying to teach one of the donkeys to respond obediently to a halter, so she can come watch the show. We're enthralled and alarmed by Terry's mighty manoeuvrings as he levels the bank, smoothing out the erosion from the last flood, creating an area where we can plant trees. Then he marches his machine to the recalcitrant she-oak, leaving it on a perilous angle, whilst he drags a huge chain around the protruding branch and begins a tug of war. I can't watch as he tries to pull the tree back against the bank – he's tilted so far he'll roll over and be killed. At one point I scream and Col just laughs. Aurora retreats to the bridge, safe from the noise of the dozer and her mother. The heavy chain snaps the branch, but the tree won't budge.

Next moment Terry is astride the trunk, in the middle of the river, pulling the cord on his chainsaw. There's a satisfying roar and he begins a circus act of balance, precision and strength. He slices one way and then another, until the blade is so deep in the trunk that it touches water, creating a geyser that drenches him. At this point Aurora and I applaud.

Then he climbs back onto the dozer as Col rearranges the chain. With tracks churning, five tonnes of machine try to shift the tree – and, despite all the sawing, it looks as though the tree will remain unmoveable. Then, with the sound of ripping and tearing apart (the kind of noise I imagine the *Titanic* made as the plates of the ship were ripped by the iceberg), Terry triumphs. Inch by inch the tree shifts. Branches splinter, the trunk complains, but the job is done. The tree is against the bank, the move creating a new place for critters of all sorts.

A few years back the official attitude to the tree would have been that it was debris and should be dragged to a sandbank in the centre of the river and burnt. But today it will be used to encourage biodiversity.

As Col, Terry and their dozer depart, I throw Haifa-white clover over the newly mounded bank. A shower of afternoon rain will be enough to get the seed germinating, which will attract the moorhen, ibis and spoonbill.

Already, down in the river, catfish and carp are investigating the rearrangements. The cormorant, little egret and pelican visit and I see a couple of turtles sunning themselves on the trunk.

I have fond memories of fishing with my Dad in a dinghy at the mouth of the Murray. Usually we'd share a pleasant silence, rods dangling. After a few catches (I can't recall leaving empty-handed) we'd row ashore and drive home, where Dad would slice the soft bellies of the fish, later burying the guts for fertiliser. Then he'd whisk a beer batter and cook the fillets, regardless of the time of day. At every meal for the next few days, including breakfast, we'd eat fish.

After fishing off a pier in Queensland with her cousins, Aurora decides she's an angler. She knows her swimming hole in the Pages is home to catfish, carp and eels, so Aurora goes fishing with her Dad, as I did with mine. Phillip is proud of her abilities. She has the knack and almost always catches a big catfish. Though, after de-hooking, it's thrown straight back into the water. She absolutely refuses to eat one.

The New South Wales Freshwater Recreational Fishing Rules and Regulations are clear: anyone wanting to fish must have a licence, except for pensioners, under-eighteens and Aborigines. Licence holder or not, there's a limit to what anyone can legally catch. The list covers all fish in all streams (regulated or unregulated) and provides a legal minimum length for every fish, plus regulations on bait, bag limit and dimensions of traps and nets. For trout fishing, even the rod, line and hooks are regulated. All of which somehow crushes the simple act of grabbing a rod and wandering down through the lucerne to the Pages with a couple of dogs for company. Anyway, if you don't catch anything, it doesn't matter, because you can still swim without a licence.

Since most local knowledge of fish comes from fishing, small-bodied fish are usually ignored or, at best, misunderstood. Although we've often glimpsed small fish in rock pools in Kewell Creek (wondering how they've survived the long dries), most of the time the only tiddlers we sight are baby catfish. Seeing catfish and eels always gives me hope. I feel a sort of pride in their existence. For despite all the challenges – weirs, alienated flood plains, levees, siltation, salinity, erosion, episodic pollution, de-snagging, water extraction during low flow, loss of natural flow ability, diseases, competition with the aliens, to name but a few of the problems fish contend with – they've survived. No mean feat when you consider historical documents reveal there were twenty-seven native species in the Hunter catchment last century. Although New South Wales Fisheries hasn't undertaken a complete survey of the whole catchment recently because it's so big, many areas were monitored during the 1990s, revealing the native species count to have dropped to twelve – a reduction of 55.6 per cent.

In the past, if we'd bothered to learn about our fish, we'd have noted those with suggestive names like bullrout, mountain galaxias, dwarf flathead gudgeon, common jollytail, yellow-eyed mullet, silver perchlet, snub-nosed garfish, large-mouth goby and small-mouthed hardyheads. Many travelled hundreds of kilometres to the estuary to spawn – a journey made increasingly difficult by dams and weirs and fords distorting the flows. Flood mitigation hasn't helped, and many of the flood plains needed for breeding no longer exist.

Now these native fish have to compete with an extensive distribution of seven introduced species: mosquito fish, carp, goldfish, rainbow trout, brown trout, silver perch and golden perch. Those little fish we'd seen in the waterholes were probably mosquito fish, which were introduced to Australia from the USA in the 1920s, in the belief they ate mosquitoes (they don't). Gone are what we called guppies, those ubiquitous little brown fish of my childhood, which were most probably empire or firetail gudgeons.

So perhaps it's not good enough to fight to keep the river as it is now, today. Some insist that the river must be re-created, helped to regain its past. Some would be content with the river they remember from childhood. Others long for the river that pre-dated white settlement. There's a dream that the river can be returned to a time when waterholes weren't filled with silt, when more trees shaded the banks, weeds weren't so prolific, nor floods as fast – when fish, other than the wretched carp, were abundant.

Do the water birds mind the dwindling of native fish?

The still water's edge is perfect for raising the young of all species and I watch a pair of masked lapwings proudly parade their young. They're not bald and ungainly, nor fluffy like ducklings, but perfect miniature lapwings, their weird yellow facial wattles dangling while the three of them, eyes to the ground, pace a patch in the shallows, as if they've lost something and just know it has to be there.

It's been years since I've seen them nest in the paddocks close by. Once I innocently stumbled upon a messy nest with absolutely no protection in a wide open area on the lawn. The adult birds' hysterical squawking slowed me just in time to avoid murder by foot. Next day I returned with binoculars so I could observe them without getting too close, but the two pale olive-green speckled eggs were gone, the nest intact but vacant and the parents nowhere to be seen. Had they managed to move the eggs somewhere safer? Had a fox taken them? Occasionally, there'll be flocks of them wading in the shallows, as if congregating for a search party for whatever it is they've lost. A few days later they will have flown away for months.

The latest, deepest drought tells a fish story that we'd never suspected. Halfway up the mountain there's a dam hacked out of almost solid rock that is always at least half full. Then, one day, Phillip noticed a strange shimmering in the middle of it: a circle of gold that he described as resembling something out of *The Lord of the Rings*. It turned out to be a few hundred goldfish (water vermin to some, but fascinating pets to Phillip) engaged in a mating ritual. Goldfish? In a dam half way up a mountain?

Then the dam began to shrink and the white-faced herons moved in. 'Well, wherever the goldfish came from they're rapidly disappearing,' Phillip said sadly.

Yet when the dam has been reduced to a muddy pool, just inches deep, he and Aurora discover it is still seething with goldfish. No matter how many the herons ate, there seemed to be more. So he and Aurora fill the truck with old sheets of iron, boards, buckets and nets and mount a rescue operation. Using the galvo and the boards to prevent sinking into the ooze, they rescue perhaps 1000 goldfish, big and little, and bring them back home, pouring them into cattle troughs and ponds around the garden. Phillip and Aurora compare their efforts to saving beached whales. Word of their heroism (or should that be madness, for saving an introduced fish?) spreads, and one evening an unidentified fax arrives containing a recipe for Creole Goldfish Stew. We haven't tried it.

When we have some rain and the Pages briefly swells again,

we find that catfish and carp aren't the only survivors. To widespread astonishment, trout are seen jumping along the river. Where did they come from? Big, fat, ready-to-eat trout are caught at Gundy. First goldfish, now trout – it is all very mysterious.

We've sold most of the cattle, a few have died in the dams, and Phillip has been out with his rifle giving the *coup de grâce* to the suffering. Now the paddocks aren't agricultural but geological.

As an orange sun dawns, I'm pacing in the olive grove, eager to experience the day while it still holds some optimism. The earth between the rows is so hard that my legs ache, as if I'm walking on cement. It's been nine months since we last needed to mow the grove. What will happen if and when rain comes? With weeds and grasses it's all a matter of timing. Each month guarantees a different germination.

As it clears the horizon, the sun reveals a luminous haze. It isn't just smoke. It's other people's topsoil and particle matter from storms and bushfires. The air is full of muck that coats everything, inside and out, on us and in us. You breathe dirt wherever you go. Fires burn across the state as every other lightning strike causes ignition of dead leaves or an old log. If I wipe the olive leaves with my fingers, they're instantly nicotined. It is this dust that dulls the brightness of the olive flowers, which are opening.

As soon as the sun hits the blossom, the vibrating hum of bees begins, whilst all the other insects, particularly the butterflies, fly chaotically as if confused by a dust-filled sky. I come here for tranquillity but, as far as the insects are concerned, it's peak hour.

At the end of a row, I notice a tree that doesn't look right. Getting closer I see it's covered in ant trails, the leaves stained by the sooty mould. By the time I drive home for some white oil spray, a wind has whipped up so strongly it's pointless going back. It blows like this for weeks. I sit on the veranda sipping tea, watching the haze thicken. Soon even the first row of hills is shrouded.

We've held the cattle back from grazing the river bank, but kangaroos and wallabies are feasting. Under the shade of an ancient rough-barked apple (*Angophora floribunda*), home to an active beehive, native tobacco (*Nicotiana suaveolens*) ignores the drought and defiantly flowers. So all is not lost. Some yellow-box trees are flowering intensely and, to my astonishment, a half-dead tree in the river paddock – an old-timer no one had been able to identify, but which many thought was a lone survivor from an ancient garden – has produced a few tiny blooms. They hang there, sad, yet brave. It's probably the tree's last hurrah and it's trying to put on a good show. Meat ants, from a huge nest underneath, climb up and down the trunk in their thousands, devouring the nectar. I rush home to collect Phillip and the Nikon, insisting that he take close-ups while I collect and press samples. He stands in the tray of the truck, focusing, framing and clicking, but it turns out the camera is empty.

Travis Peake, the botanist mapping 'remnant vegetation'

for the Hunter Catchment Management Trust, confirms from a pressed sample that the dying tree is, or was, a western rosewood (*Alectryon oleifolius,* subspecies *elongatus)*, a small tree not usually found this far east in New South Wales.

'It's at the very limit of its range,' says Travis. And at the end of its tether.

'How old could it be?'

'I'm not sure, but it could be up to 500 years.'

So here, overlooking the Pages River, our unlovely but tenacious rosewood – only half alive on one side with a few melancholy tufts of diseased leaves, yet still maintaining a domed canopy that hints at its former glory – is one of the brave hearts that forged ahead, seeking new frontiers centuries ago, when the climate was drier and colder. Now they're retreating back to their main habitat west of Dubbo.

Rarity of any kind is attractive. We have only one western rosewood, which has been all but ringbarked from years of cattle scratching themselves on its trunk, and only one patch of white native tobacco and one rare cluster of sandalwood trees.

Travis thrills to his job, one of the most important, I reckon, in the Hunter Valley. By mapping the last of our native vegetation he's creating an important point of reference for our future understanding. How much I appreciate having this enthusiast to talk to about plants. In years to come, I'll look back and say, 'I knew Travis when . . .'

Travis says there are possibly only fifty *Cymbidium canaliculatum* left in the whole of the Hunter Valley. If he's right, then

Elmswood has almost half the surviving population. These tiger orchids, often found wedged in the armpits of dead trees, have always been a springtime delight for us. I invite Travis over to conduct a *Cymbidium* census. He arrives with a truck loaded with cameras, lenses, plastic bags for samples and boxes of books to help us identify plants as we go. I keep a tally of everything we observe in my notebook, not just the orchids. I write down the correct botanic names and the families of various plants, bushes and trees, so that I'll be able to locate them in my books.

We find twenty-six *Cymbidiums,* attached to three different types of trees, and now they're all recorded and numbered with white cattle tags nailed to their trees.

13

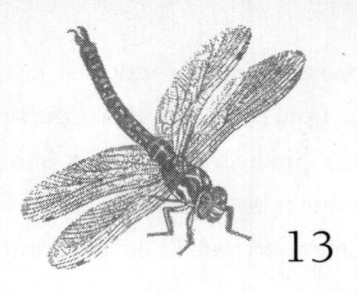

The public liability debacle, triggered by the HIH collapse, arrives in Gundy in 2002 when our Family Fun Day, an annual event held on the last day of the Scone Horse Festival, is cancelled. The community is devastated, most of all the children.

To make a political statement, Mick from Gundy builds a big coffin and Judy wraps it in black fabric and paints R.I.P. on it. Then we tie it to a trolley, and Steve and John put on their dickie-suits, so they'll look like undertakers as they wheel it in the Scone Horse Parade. Aurora and the other Gundy kids, dressed in black, walk beside the coffin, their sandwich boards proclaiming 'INSURANCE KILLED OUR FAMILY FUN DAY'. The funeral cortege is loudly applauded as they march down Kelly Street, in between school floats and clip-clopping ponies.

Being the optimistic place it is, Gundy is not dissuaded from going ahead and mustering voluntary labour to build a new amenities block at the rodeo ground: an open pole shed with a galvo roof and a kitchen area. Before it is finished, locals are already planning a new event to show it off, a Gourmet Food Fair.

One part of this fair will owe its existence to an old Family Fun Day tradition – the Fun Run, the first event of every gala day, in recent years attracting loads of entrants, not just locals but semi-professional fun runners from far and wide. Yet the problem with public liability and narrow public roads remains, and we get a phone call asking if the run might be re-routed so as to cross the Pages and come through Elmswood. An 11-kilometre route is proposed. Of course, we agree.

But the drought is deepening and Elmswood is looking terrible, pitiful, parched, miserable. That's true of the entire district, so I suggest this year's event should be cancelled, or at least delayed until the rains come. Who'd want to come to Gundy when everything is dust and heat and despair? Fortunately, no one listens. Plans are afoot, enthusiasm prevails.

The day before the event I ask Phillip to drive over the track and remove any dead carcasses that may be visible to passing joggers. There are, of course, carcasses they can't see, but may still be able to smell. I know of three cows that have died in a dam further up the gully and later been dragged near the track, probably by feral pigs. Although they probably aren't pongy any more, they're not a pretty sight. So Phillip grabs a length of rope and heads off in his ATV, returning an hour later shaking his head in wry amusement.

'One cow that bogged in the dam took me twenty minutes to drag a few yards behind the 4WD truck in low gear,' he says. 'It was that bloody heavy. And today, there it is, still looking like a cow. But guess what? All its innards have been removed by eagles, goannas, foxes, crows, ants, you name it. I could just pick it up and carry it. So I threw it behind the hill.'

In due course, the fun runners run through Elmswood (all the corpses safely out of sight) and the inaugural Gundy Gourmet Food Fair is a huge success.

The founding fathers of this land of sweeping plains failed to provide the Commonwealth with direct power over any aspect of the environment. So it wasn't until the 1990s that Canberra finally began to enshrine an environmental aesthetic in law. The government discovered that it had the power – and the right – to initiate debate and to coordinate various 'national strategies' for Greenhouse, Ocean Policy, Forests, Ecologically Sustainable Development and the Conservation of Australia's Biological Diversity, for example. The National Action Plan for Salinity and Water Quality was gazetted in 2000. The Council of Australian Governments' (COAG) Water Reform Framework kicked off the reform ball, but the Commonwealth needed a stronger legal structure to ensure its implementation – so a new act, the *Environmental Protection and Biodiversity Conservation Act*, 1999, was passed by the Reps and Senate, and approved by the governor-general. In December 2000 the *(NSW) Water Management Act* was passed. Here was the law that would 'provide for the protection, conservation and ecologically sustainable development of the water sources of the State for the benefit of present and future generations'.

So there is a way, but is there a will? Is there still a BHAG (a Big Hairy Audacious Goal)? After two years of discussion, our water management plans have been reduced to a mere water sharing plan. All the other BHAGs have been dumped in the too-hard basket. Once the debate about 'best practice' began, it became clear water use was really a land-use issue and it was way outside the DLWC's and the Committee's frameworks.

As far as I was concerned, the BHAGs were the only exciting part of this water reform, like reading about the commitment to abolish slavery or colonialism in centuries past. The divvying up of entitlements was the boring part, yet here we were once again not addressing the crux of the issue. The BHAGs were being buried. I felt like I'd been walking down a road, and suddenly discovered that someone had changed the signpost and hadn't told me.

New frameworks and the reshuffling of departments are proposed. Implementation of blueprints and laws are still supposedly going to happen. Having a law doesn't mean that it'll be implemented or that the response will be anything more than tentative. Tim Bonyhady showed this clearly in his beautiful book, *The Colonial Earth*. Our early governors pronounced laws to protect various things (like banana trees, the Tank Stream, birds, seals, river banks and more), but that didn't necessarily change, or guarantee protection for, anything. Even in the colonial era, the powers-that-be succumbed to electoral pressure and the demands of the privileged.

More challenging is the fact that today's landholders who do good works may not live long enough to see any positive consequences. The uncertainty of outcomes is the main obstacle to engendering enthusiasm.

Yearning to be part of the solution and not a contributor to future problems, I've always been keen to do something that, in the long term, will lead to a better farm and better farming. Friends and enemies alike point out that I 'waste' a lot of energy on what are essentially 'unproductive' occupations like tree planting, fencing off, slashing, restoring unused out-buildings. I'm the sort of landholder who earnestly reads documents headlined 'WHAT CAN YOU DO?' Yet even I am not always convinced by the assumptions they express and not entirely persuaded by the latest theory.

Of course, not having conclusive 'scientific evidence' shouldn't be an excuse for doing nothing, which is what people have done for decades. History reveals that, even when our rivers have verged on the tragic and the terminal, not much has happened. Generations have tended to pass a torch that was more smoke than fire. At the very least we have a responsibility to ignite it and run, like Cathy Freeman, towards the cauldron. Passing the torch is far, far better than passing the buck.

You don't have to be Einstein to calculate that carving up land and selling it bit by bit will generate more dollars than selling it as a whole paddock. White Australians have been subdividing land ever since they took hold of it, and for most farmers these days the only fat cheque they'll receive is the one they get when they walk away. That's why subdivision is called 'the last crop'. In the 1990s Scone Council's Development Committee recognised the need to protect good agricultural lands and had gazetted in our Local Environment Plan (universally known as the LEP) stretches that weren't to be hacked into 40-hectare blocks.

This was when philosophies of land use, many initiated at the 1992 Earth Summit in Rio de Janeiro, were beginning to influence policies around the world – even in Scone. It was also at this time that my friend Stuart, a conscientious environmentalist, gave me a folder outlining the idea of Local Agenda 21 – a process of drawing up and implementing local ideas for sustainable development. He suggested Scone Council should have such a committee to encourage sustainable policies. Local Agenda 21 felt like our BHAG. It stated that if change was inevitable, given the irresistibility of free development, then it had to be as harmonious and undamaging as possible; moreover, it had to begin locally. Change was, first and foremost, a municipal matter.

By the year 2000 Scone Council established its Local Agenda 21 committee, under the chairmanship of Peter Hodges, the Scone councillor who'd been elected on his environmental credentials, acquired after years of campaigning against the Parkville Piggery. At one of the early meetings he observed it was tragic that old dead trees and aged single specimens of eucalypts were being bulldozed and burnt to make way for yet more development. He talked about how certain birds need certain trees and posed the rhetorical question, 'Why should a new landowner be able to just come in and wipe out old trees important to an area?'

But how would a new owner know which trees were important, not having watched the migration of birds over decades?

In any case, old trees, especially common-or-garden eucalypts, are rarely seen as significant. Peter suggested Scone Council should have a *Tree Preservation Act* to help it keep an eye on things; the committee agreed. At our next meeting, we look at the *Native Vegetation Act* and discover it's largely constructed around loopholes. You could have been forgiven for thinking it was an act designed to allow wanton clearing of almost anything with roots. The more I learn of the exemptions to the regulations, the more my eyes glaze over as I imagine Local Agenda 21 trying to push more regulations when everyone in the shire is sick to death of the ones they already have.

Take, for example, recent new rules regarding septic tanks. Most rural people considered this latest obsession with dunnies as a poor excuse to put another inspector on the council payroll. Why focus on the flush when there were significantly more important issues at stake? Like tree clearing, salinity in the Hunter, sulphur emissions from the power stations and energy-efficient street lighting.

Just as I'm slipping into depressed thoughts, Stuart says in a loud voice, 'Please, no more rules. What we need, what we want, are people to value trees, to see them for what they truly are: not just objects of beauty or objects that get in the way, but as something important to our biodiversity. To our whole life.'

And I agree.

The *Australian*, 30 July 1828, mentions the sighting of a volcano north of the Segenhoe village. In 1829, Rev. C.P.N. Wilton, chaplain of the Female Orphan Institution, Parramatta, was carrying out geological investigations, and discovered a smoking crevice. When he threw stones into the abyss, it seemed bottomless. Today we know that 'volcano' as Burning Mountain, a burning coal seam north-east of Wingen, an Aboriginal word meaning 'fire'.

Back then the 'volcano' was 150 metres from its present site and emitted both a flame and a bluish column of sulphur with black tar (a sort of bitumen) oozing along the edges of the cliffs. Rev. Wilton claimed the surrounding sulphurous and aluminous minerals were already being successfully applied to cure scabby sheep. Liquid potions and ointments, supposedly with medicinal value, have long been made in the area and were being sold as late as the 1960s.

When I visit, a very light puff of smoke from a vent is all that's visible. The temperature at ground level is 350 degrees Celsius and the burning coal seam itself as high as 1700 degrees.

It's one of only three naturally burning coal seams in the world and is now a nature reserve managed by the National Parks and Wildlife Service. Its significance is upheld by the Australian Heritage Commission, the National Trust and the Geological Society of Australia.

Tom Miller, a Wonnarua descendant, tells his people's story of Burning Mountain and a nearby rock formation called 'The Wingen Maid'. A group of warriors set out to do battle with a neighbouring tribe, leaving behind their old people, women, and children. At the end of the day, one warrior didn't return. His girl sat waiting for him, praying to the sky god Biami, who, impressed by her sorrow, turned her into a vast sculpture, The Wingen Maid. It was her burning tears that ignited the coal that burns forever. (What actually started the coal seam burning some 5000 years ago is uncertain. Perhaps it was a lightning strike or spontaneous combustion – which is possible, given the abundance of sulphur in the region.)

While the actual site of the burning coal is being preserved today, part of the burnt seam formed kaolite (or flint clays) and mullite and was mined in the late 20th century. Flint clay is a natural chamotte used to manufacture firebricks and other refractories. The quarries high up in the hills on the western side of the Pages are now deserted, filled with water and surrounded by a tall fence with Keep Out signs posted. I fear this history could be buried with overburden from the Bickham Coal Mine.

Many of the great cities and regional towns in Australia owe their existence and their magnificence to mining. Not just Bendigo, Ballarat and Broken Hill, but hundreds of towns across Australia depend on their historical connections with mining to celebrate their existence. That's the case in our part of the world with Cessnock, Kurri Kurri and Nundle. Even tiny Stewarts Brook, east of Gundy, boasts a mining past – 150 years ago, gold nuggets littered the brook and hundreds of families lived in the narrow valley.

Although Murrurundi was originally settled for cattle grazing and later profited from wool production, it was a mining venture that boosted its economy when oil-shale rock was discovered in 1860 at Temi, just north of town. Sometimes called kerosene shale, it's part of a coal seam. When heat is applied to the rock, oil can be extracted, which is then made into various types of liquid fuel, such as kero.

When the Club of Rome warned the world of our depleting resources, the dissident scientist Sir John Maddox pointed out that there was enough oil-shale rock in the world to sustain our profligate automotive industries for the next 14 000 years. Even though it is almost as hard to get oil from a rock as it is blood from a stone, shale has been an invaluable resource for countries denied oil due to war or embargoes. It's an industry that has grown and died at various stages in Australia's history – a bit like olives.

By 1901, 500 tons of shale had been mined, but that was nothing compared to what was about to happen. A syndicate was set up in 1905, which by 1910 had raised £300 000 to take the shale from Temi and convey it, via an overhead cable system, over Mount Page (now named Mount Helen), 1769 feet above the mine site. It was then railed to Hamilton, near Newcastle, where a newly constructed refinery produced kerosene and lubricating oils. The engineering this operation entailed was as remarkable in its way as the construction of the Eiffel Tower or the Panama Canal. Moreover, overnight, little Murrurundi acquired 200 jobs.

During World War I the mine encountered engineering and financial problems and lost labour to the armed services. It didn't help that it was partly owned by Germans. When peace arrived in 1918, the refinery finally went broke.

The huge buildings that once dominated the north of Murrurundi have long since been dismantled, with any hint of brick and cement consumed by blackberries. Now the promise of a mine is exciting some of the people of Murrurundi, at a time when the future of coal seems as dubious as the future of oil shale.

Thousands of large orange-black wanderer butterflies (*Danaus plexippus plexippus* – sometimes called monarchs) are strobing an entire valley. Their fluttering is so dense, so confusing and fast, that it's like entering a disco. It makes you want to dance. The butterflies rest on cotton-wool bushes (*Gomphocarpus fruticosus*), a straggly upright weed that pops up after rain and flourishes in earth turned over by wild pigs. Sometimes butterflies can be found all over them, like flowers. *I'll pose here for a minute, does this look good?* They move along the stalks and relaunch themselves into the swirling celebrations. The evolve from juicy, striped, fat-as-your-finger caterpillars. Not natives, they supposedly arrived in the late 19th century, along with the plants colonials introduced for their dietary requirements. Since my garden doesn't offer the right menu, I rarely see them around the house, but in the hills and valleys along the river they are everywhere.

Our most charming is the huge native orchard butterfly, *Papilio aegeus*, its wingspan up to 120 millimetres. Their caterpillars, which resemble bird droppings, strip the leaves from the bush lemon on the patio until they're so fat, so stuffed, they're unable to move, let alone find somewhere to hide. Then they begin to cloud over with the first layer of cover, their chrysalises attaching themselves to all sorts of places: the winter wood stack, the brick wall, laundry cupboards and under surviving leaves. Later, when the butterflies emerge, the patio is full of their fluorescent wings.

There's been a great increase in the variety of butterflies and moths over the years. When I find their vivid corpses outside, I place them very carefully on top of a vase full of dried carrot seed heads. Visitors look at them curiously, unsure if they're alive or just resting.

When we were negotiating the purchase of Elmswood in 1986, John Wilson was finalising his move to Gunnedah, to be a New South Wales Department of Agriculture environment officer. One of his duties would be to serve as a representative on the North-West Catchment Management Committee. I got to know John when the Scone Council reassessed its rural land development plans in the mid 1990s.

John would go to local meetings and, in his Irish accent, tell the truth about the value of rural land. He'd worked in the western

Sydney region and had conducted what he described as a 'windscreen survey', where he actually drove around the district to meet the farmers and add up the value of their produce. This led to the revelation that agricultural business in the Sydney basin alone was worth not the miserly $460 million that the bureaucracy had attested, but well over $1 billion – meaning that the remaining rural land had been grossly undervalued by officialdom. How could I not like him? John was passionately defending agriculture as a valuable part of society, and had facts and figures to back his arguments.

His job, to defend the bleeding obvious in New South Wales, seemed a long way from his background and training in Ireland. After studying marine zoology at Dublin University, he'd worked as a writer/editor at the influential *Nature* magazine in London, where he was contacted by the New South Wales Department of Agriculture seeking help in getting international recognition for their scientists' work. Marine zoology involves water, albeit salt water; and since water is the key ingredient for agriculture in this driest of continents, John's new career makes some sort of sense. Little wonder he finds himself on so many committees.

John calls in for chat. We sip tea and I ask, 'What committees are you on anyway?' There's so many he can't quite remember. Change needs dialogue, that I can understand, but the committee network seems cumbersome and abstract. Too many people talking water reform that from a 'down on the farm' perspective seems merely to threaten wasting time and paper. I recall the warning of a sceptic: 'Conservationists are just conversationalists.'

The Natural Resource Management Committee for the Hunter region alone has ten subcommittees, each holding monthly meetings. New South Wales, divvied up into seven regions, has spawned fifty-seven committees – all talking the same stuff, just variations on a theme. When you break them down and include all the local meetings going on in memorial halls, you'd think the entire rural sector would be abreast of the issue by now. But as John patiently points out, 'the level of complexity has risen dramatically over the past two years.' (And, dare I say, the level of confusion.)

'Knowledge is labyrinthine,' smiles John. 'And sometimes committees don't represent groups at a deep enough level.'

Perhaps we'll achieve that once we've digested all the information and endured meeting fatigue. While I'm staggered by the breadth of the debate, by the complex infrastructure now set up across the nation, I still wonder whether the task is impossible; and, if I were in charge, where would I start? What would I do? Yes, meetings are a benchmark of democracy; it's just a pity there's no precedent around to help us.

A depressing study, *Australian Farmers' Attitudes to Rural Environmental Issues 1991–2001*, reveals there's been virtually no change in attitude after a decade of Landcare. While environmental issues may be better understood, there's an overall decrease in concern about the 'seriousness of land degradation' and less support for conservation organisations and their activities. If people don't understand what's going on, it's not because the information isn't out there; it's because people choose to ignore it.

And there's something I've learnt about meetings, having

gone to so many on water issues that have ended in anger and disagreement: meetings are rarely meetings of minds. They're battlefields of conflicting agendas. But I'm still glad they're happening.

※

There's something perverse about paying for the repairs on our nation's rivers by selling the electronic rivers of Telstra. As soon as publicly-owned Telstra is partially dumped in the free market, our rural telephone service begins to deteriorate. Despite being required to pay extra for 'business' services, those of us who live in the bush are grossly disadvantaged. Costs soar, response times stretch, services 'cease to flow' and become ever more expensive. Mobile reception? Largely a joke. The promised benefits of the Digital Revolution are just something you read about in the papers.

So a national asset is sold off, on the grounds that our rivers will be repaired. To add insult to injury, any economist (in fact anyone who can add up) realises that the revenue from the sale could never begin to do the job. Rivers, the water and their banks have to be regarded as public property. I'd like to see all the river banks returned to Crown control.

Instead, we're told that the 'free market' and private ownership both care more and are wondrously more efficient. Yet it was the clearing of private land along the Pages that created the high-energy stream that the river is today. Moreover, across the country, private projects of various magnitude (including the thirsty rice industry and the chemically profligate production of cotton) will create immense ecological problems that, down the track, downstream, will have to be tackled with taxpayers' funds.

The private/public contradictions abound. We're told that we must have the coal industry to pay for environmental works, particularly those along the rivers. Yet the coal industry creates much of the environmental problems in the first place. Where most mines are now clustered between Singleton and Muswellbrook, the gouging-out of coal has interfered with aquifers and gravel streams that can never be replaced. The overburden – reshaped to look 'natural' – is made of huge rocks and clay soils that can, in a good season, with loads of fertiliser, produce a thin sheen of grass. But as we've observed throughout the valley, after a few bad years, the trees die and the grasses disappear. Why are we trying to disguise the fact that mining ever took place? Why waste all that money and effort? 'Coalmines are like feral animals,' says a scientist, 'they're trying to disguise their scent.'

It's painfully clear the true cost of economic growth has to include the deterioration of the environment, and until it does our accounting system should rightly be considered a failure and a con.

14

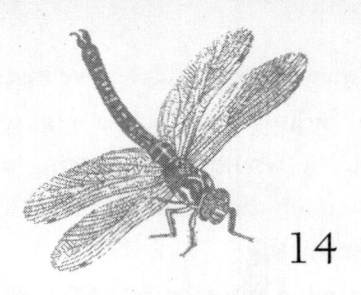

The Australian olive industry is heading where I suspected it would, where agriculture and business in general around the world seem to be heading. Consequently, hundreds of olive groves across the nation are being neglected. People I knew in 1997, when we were planting our first trees, have already given up, sold their farms and moved on. Others have taken higher risks and plunged deeper into the industry and investment, planting more trees, buying presses, opening tourist outlets. Significantly, Olives Australia, the company that kick-started the olive industry in the mid 1990s by selling 4.5 million trees, has closed their nursery.

The learning curve, for all of us, has been steep. Each year the Australian Olive Association and numerous regional groups have held meetings bringing together the great *oliviers* from around the world. Local growers travel, read every available book and prowl the Internet. Consequently our industry's collective knowledge is, to use the clichés, 'cutting edge' and 'best practice'. Whilst the rate of tree planting has slowed, the big groves are getting bigger and the alliances with multinationals are already under way, echoing a sequence of events we've observed in the wine industry.

It's estimated that Australia now has 8.5 million olive trees. If all were productive, this could mean we'd supply a mighty 1 per cent of global production. We would be producing around 30 000 tonnes of oil, pretty much the same volume as we currently import to meet domestic needs.

The majority of trees are in big groves of 100 000 to 500 000 trees. Victoria leads with 27 per cent, leaving New South Wales, Western Australia and South Australia trying to catch up. Many consultants, however, think that the tree count should be reduced by 50 per cent, because too many trees are unproductive, due to lack of pollination, and many small groves are being deserted.

As we walk through our grove examining the leaves, squeezing the olives, calculating ripeness for hand-picking, the big groves are increasingly investing in mechanical harvesters. The idea of overhead machines as big as houses clumping through the groves, raking or brushing off fruit seemed fanciful just four years ago. Now it's standard practice with the bigger businesses. We know we can grow olives and process them, but the cost of harvesting could still be the industry's downfall – as was the case during an experimental flurry in the 1960s.

After the 2001 biodynamic AGM at Merrigum in Victoria, I drove to Boort, passing endless irrigation channels dug in the 1930s, to meet Ken McDougall who was, at the time, manager of Australia's largest olive grove, which is run by Olivecorp. The flat land, appropriated by settlers in the 19th century for sheep and wheat production, is a Cubist's dream. All roads are laid out on grids, so that diagonal driving is impossible. You have to take endless left and right turns. The mazelike landscape disoriented me for most of the journey. Eventually I found Boort (population 800, average rainfall 330 millimetres) identified by a sign (since removed) boasting it's the 'Oasis of the North'. Such efforts in tourism marketing always make me smile. Why not let a landscape simply reveal itself? Why does it need a slogan or a caption?

I drive through town in search of a coffee shop and, yes, Boort needs something, I'm not sure what exactly, but perhaps olives will do. Boort already produces 40 per cent of Australia's processed tomatoes. Not that you'd notice. Tomato production is intensive – since the plants yield 85 tonnes to the hectare, you don't need many hectares to dominate the market. To the modern agriculturalist, we're dealing with a proud statistic. But it's another example of technology pushing production to the limits of the land and commonsense. Also, it makes the growers vulnerable. They'll be out of business if processors drop their prices any lower – and Australians continue to buy cheap imported products.

We all know that most fresh tomatoes don't taste like they used to. Why would they? How could they when they're bloated with artificial fertiliser? Tomato – that wonderful food that goes so well with olive oil and salt – has now joined the soya bean and canola as a victim of genetic engineering.

In 1998 the US Food and Drug Administration (FDA) was taken to court by the Alliance for Bio-Integrity, a non-profit coalition of concerned citizens, scientists and religious groups of all denominations. They hoped to prove that GE food was unsafe, and that the FDA's claim that scientists unanimously supported genetic engineering was untrue. The case revealed extraordinary research data, in particular concerning genetically engineered tomatoes. Unfortunately, the fruit caused stomach lesions in mice in a number of research projects. This was bad enough, but worse was the fact that the data was suppressed. Nor did the FDA consider these findings 'significant' during its assessment process. Two and a half years and 44 000 pages of conflicting data later, the US District Court in Washington DC upheld the FDA claim that GE foods aren't materially different to conventional food. It's all about risk assessment and, in this case, the degree of risk from GE food was deemed acceptable – as the risks of DDT once were. But as far as I'm concerned, tomatoes like that aren't for eating. They're for throwing.

Thankfully, GE tomatoes haven't made it to Boort yet. However, there are some serious environmental issues that the area has to face. Like the desolation of the landscape caused by dead trees. You see arboreal death on a similar scale in the New England area north of us, where dieback has been caused by clearing, which has dramatically affected the population of flora and fauna, encouraging the spread of hungry bugs. Around Boort, trees die because of salt.

After settlers cleared the area and built their houses, the lakes began to dry up. Soon the white sheen of salt began to appear. Little wonder that salt is a principal issue concerning Ken McDougall. 'In 1992,' he tells me, 'I ran a workshop called "using saline groundwater for profit" and 150 people came.' The soil in many of the paddocks around Boort is now saltier than the sea, despite the last oceanic inundation being thousands of years ago. However, Ken believes there are engineering solutions. 'We have to treat salt as a "good idea".' That's the thing about Ken. He's a perennial optimist, the only person I've met who smiles when he talks about Australia's salinity crisis.

Everywhere I drive I see the effects. Everywhere are saltpans and dead trees. Yet Olivecorp, a subsidiary of Timbercorp, decided to plant Australia's biggest olive grove here on the 'unsalty' side of Boort, prompting a new slogan to be launched in the region, 'The New Mediterranean'. What's Italian for *chutzpah*?

Meet 100 per cent corporate agriculture. In 1999, Timbercorp, a public company, sets up a new structure called Olivecorp. It raises millions of dollars from city-based investors, mostly individuals seeking tax write-offs, and buys up land around Boort, naming sections of their project after those who'd sold the land. They start planting what will be 3000 hectares of olives, or about a million trees.

'This is as flash as it gets,' says Ken as we drive around. 'Nothing has been left to chance.' But I was pleased to learn that hares chew corporate irrigation pipes, not just Elmswood's.

In a portable office, its walls covered with maps and diagrams, two blokes sit at computer terminals checking irrigation and fertigation schedules for the whole grove. Eventually the olives will receive as much water as a lucerne crop (at least five and a half megalitres per hectare). Such computerised irrigation systems are often used as proof of 'sustainable' water management. I confess to feeling a pang of envy for the monitoring gear, knowing that it's far too expensive for a grove of our size. I also believe that having the gear is only half the equation. Knowing how to read and analyse the data properly must be dependent on your horticultural and soil knowledge. And to acquire that, surely, you must get your hands dirty.

When I see the soil these olives are planted in, surrounded by bare, brown, herbicided earth, it's no wonder that the small trees are being watered every two days. The red brown soil, if that's what you could call it, looks more like dust than earth. I wonder if it will ever be possible to build a healthy new texture, with enough organic matter, to sustain an olive crop. Or will it always be drug dependent – on the liquid drip of fertilisers flushed into the irrigation system?

According to Ken, Boort's degraded land forced the scientists and farmers to confront what was going on. 'Farmers doing the same old stuff' was, according to Ken, 'a waste of time.' The new way, the Timbercorp way, was *the* way to go. This is only possible now that water and land are treated separately. Olivecorp has bought 15 000 megalitres of water from the Goulburn–Murray Water Authority system, part of the Murray–Darling catchment.

When a company seeks to be the biggest, redefines a township and inadvertently renames a region, one can only hope it succeeds – because the scale of failure would be catastrophic. If all this money, expertise and alleged efficiencies fail, what will it mean for the rest of us in the olive industry? Is there a place for small growers in the scheme of things? Does the future have room for us? In the short term, agriculture will continue to be subsidised by small growers who value more than the bottom line. Their success stories – and there are many – are what keep my hopes high. Like the time I bought a bottle of olive oil, a brand I'd not seen before, from a small deli in a small town and discovered it was sensational. Later I learn that its proud producers have a few thousand trees, their own small press and supply their local region. These are the romantics who are developing the olive oil and fruit culture of Australia. While the big groves employ more people, support the chemical industry and can afford to embrace the latest technologies, it is the smaller grower who is providing 'olive culture', without which the domestic industry is doomed. Other romantics are pickling different varieties and creating new marinades. Consumers will be able to choose from many different types of pickled olives in the future, a fact that is certain to lift the abysmal statistic that only one in three Australians eats an olive a day. The pickled olive market in Australia is a small one, but surely if we can celebrate salty Vegemite, a waste product of the brewing industry, a good olive shouldn't be too impossible to sell.

The big keep getting bigger and the small struggle to survive. Every time I pass a 'For Sale' sign in front of an olive grove, my heart sinks. Olive growers across Australia share a dream – which may prove to have been a delusion. When I push my trolley down the supermarket aisle, I can see twenty-five different olive brands selling at cheaper prices than it costs for small-grove production, hence the 'For Sale' signs.

The cost of production versus the price of food at the supermarket is so out of kilter, so immoral, that it's a wonder anyone wants to grow anything. One top Australian brand, Viva, is selling for below the wholesale price I was offered months ago. Planning to become the biggest Australian olive-oil brand, Viva was caught blending Australian oil with imported oil. Rumours swept the industry and were met with shock, horror and, inevitably, a desire to hush up the scandal. How would the Australian consumer, whom we're praying will be ready to pay more for Australian extra virgin olive oil than for cheap imports, react to the news that the local stuff had been mixed? So much for truth in labelling. In an open letter to the industry, Viva claimed they'd been 'planning for some time to change the back label'.

It's almost as hard to get space on the supermarket shelf as it is to squeeze a camel through the eye of a needle. Origin was a company that grew out of Gwydir Grove, a brand started by two energetic women from Moree back in 1995, before Australia went olive-mad. When I met them, they'd set up a portable press in a ship's container and were towing it around eastern Australia – to any olive grove that had trees with fruit on them. Three years later they were part of a team that invested in a one-tonne press permanently housed in Inverell. Next, they sold their brand to a new company. Two years later, exhausted by the struggle for market share, Gwydir Grove and Viva merged. This 'corporate consolidation' so early in the industry's development reveals the flaws in our original budgeting. Many respond with a shrug, saying 'reality had to bite some time'.

Bite? Sometimes reality mauls. That's what happened with the news the European Union had decided to give their 2.2 million olive growers a further 2.25 billion Euros a year, over the next three years. So much for free trade. How can we create an import-replacement industry (let alone an export one) when we have to compete with such massively subsidised production? All business is tough, but you don't need an olive press to be under pressure in the olive business.

Back in 1996, when we were considering planting olives, we believed the EU would, far from increasing subsidies, wind them back. Little did we suspect that olives would represent the last stand of Europe's underwritten agriculture. We shouldn't be too pious, as our industry enjoys modest help from three tiers of government. Prospectus-based groves mean the tax office forgoes millions of dollars in revenue from big income earners. Some large groves have convinced state governments to invest in infrastructure to assist development; and councils have picked up some bills for processors.

Nonetheless, EU subsidies permit overseas olive oil to be sold – dumped – in Australia at prices that are impossible to match. There is a federal government regulation called the 'countervailing tariff' to provide protection for Australian goods at risk due to subsidised competition. Australian processors, led by the Barkworth Group, are trying to have this tariff implemented for the olive industry. It's a move that would see the price of olive oil on supermarket shelves substantially rise. Consequently, it is being met with mixed reactions. I'm in favour of it.

Why do we have to keep destroying our industries, manufacturing or agricultural? Like all decent, healthy, quality foods, olive oil is expensive to produce, and we all may as well get used to that fact.

15

Stretching in front of the homestead, all the way to the river, is our lucerne plot. Some of the paddock was sacrificed to olive planting, but the rest is sacrosanct. Lucerne looks beautiful. It is perennial, prolific and its pollen seems to intoxicate bees. Then there's the bonus that it's high in protein, deep in root and within one season from the first watering in late winter will shoot a lime-green leaf that changes into rich emerald before flowering. And when you cut it for hay or silage, it'll do it all over again, possibly five more times before the next winter. Lucerne hay keeps its colour and its fresh aroma for years and years. Even better is the smell of lucerne silage, made into huge bales mummified in green wrappings. Phillip says the bales smell like Old Holborn, the sweet, moist tobacco he made rollies from until Aurora and I stopped him. I reckon it smells like molasses.

Friends David and Heidi milk eighty magnificent cows that teeter on 100 of the steepest hectares in the Great Dividing Range. Their yoghurt and fresh cheese has been described by food scientists as the best cultured food made in Australia. Visiting their farm years ago, before the dairy was set up, when they made their living growing herbs, we'd sit at their kitchen table drinking the sweet milk David had collected that day and taste-test Heidi's experimental products. 'One day,' he'd say gently, as we spooned more of Heidi's yoghurt into our bowls, 'we're going to go commercial.'

Heidi was a botanist and David an accountant in their previous lives; now they both look as suited to their farm as the Amish to the Corn Belt. With her black plaits and rosy cheeks, Heidi could have stepped straight out of a Swiss Alps tourism poster, whilst David could have been the model farmer for Tolstoy – all wild hair, beard and big strong hands. In between growing, drying, packaging and selling herbs, they've built a small dairy and factory and put together their Ayrshire herd. Despite an average 80-inch rainfall, they still needed to buy hay to supplement the winter feed and ensure the cows could always provide their sweet milk. We started to help out with Elmswood's lucerne hay and silage. I enjoyed knowing our lucerne was being eaten by such beloved cows; whilst David and Heidi, enthusiastically biodynamic, enjoyed knowing the hay and silage had impeccable credentials.

There's something problematic about it, though. To keep their certification, they have to buy certified A-grade biodynamic hay and, since there is none to buy in their region, are forced to absorb the transport costs of buying ours. And here we were taking water from our unregulated stream to grow lucerne for a distant herd to make milk and yoghurt sold primarily in Melbourne and Sydney.

As the demand quickly grew, they had no hope of meeting it, and the costs of transporting feed and finished products meant the profit margins were very slender. It is a dilemma familiar to many small farm enterprises. Some find an equilibrium where the land's capacity sustains both the social and the economic needs of a family. Given David and Heidi's remote acreage, its herd and dairy, there was an absolute limit to growth. Accepting this, they show what sustainability should mean. They know the capacity of their land and their cows. Whilst David and Heidi are under constant strain – the workload is immense – the herd and the land are not.

We share problems as well as ideals. Having watched their business grow to its upper limit, I know that the greatest risk they face isn't the quality of feed, but the last straw of paperwork. Every three months, 'suits' visit the dairy to check on procedures. Not once have they tasted the yoghurt or cheeses or walked over the paddocks. All they ever want to see are the forms. Are they filled out correctly? In triplicate? David and Heidi are very proud of what they do, of what they've achieved, and find it insulting that bureaucrats are interested only in the documentation.

Small farms aren't necessarily good farms – especially those managed by owners who once had sizeable properties. Having been forced to sell off their land, paddock by paddock, they've a tendency to expect too much from their dwindling acreage. A 1955 report, *The Contribution of Historical Factors to the Present Erosion Condition of the Upper Hunter River,* by soil conservationist N.H. Monteith, passionately argues that farmer exploitation in our area and the refusal to accept the limitations of the land, has been the single biggest cause of problems with our rivers. The soft ground described by the first settlers was, within a few years of white settlement, a treeless hard pan, the native grasses eaten almost into extinction. And the rocks I trip over as I climb Black Mountain would have been covered in a soft tilth before the sheep arrived. Australia needs more farmers like David and Heidi.

Fewer and fewer cattle producers around here make their own hay any more. That's because they buy from Lloyd Rossington and his family, who operate a lucerne-chaff farm on the New England Highway, north of Parkville, on the Kingdon Ponds. Born in the district seven decades ago, Lloyd left school early, broke horses, fenced many a paddock, built stock yards, installed irrigation systems, share-farmed a dairy and bought his first 13 acres, for £1000, back in 1958.

'I do things in fours,' says Lloyd. 'We have four children. Four inches of rain grows a crop of lucerne. We work sixteen 4-hectare blocks of lucerne. We cut hay every four weeks in summer. We use 4 megalitres of water per hectare. We make forty bales of hay per acre. We change machinery every four years, or every 4000 hours. You have to be mathematical. Our golden rule is to make 100 per cent on what we do.'

Lloyd lectures me on the correct attitude towards farm

equipment. 'Always buy the best, regardless of price.' However, even if Lloyd could afford to buy the biggest pump on the market, he'd never dream of it. He wants to be able to water all through summer. He wants his irrigation season to be spread out, extended. A big pump would exhaust his wells too quickly and affect the long-term supply.

We're stingier than Lloyd when it comes to equipment, and for years I've put off buying an irrigator. That's one of those giant sprinklers that inch across a paddock. For years we considered buying one, but could never come at the price. I'd had sales reps explain the different styles, sizes, brands, and been all but buried in brochures. Then, just when I was about to write out a cheque, Lloyd helped me come to my senses. I'd never have forgiven myself if I'd kept up with the Joneses. Irrigators are, after all, a symbol of the profligate water use I despise. On one day in Segenhoe twenty-one travelling irrigators pumping across a horse stud ensured there was no water to flush the toilets in the adjacent youth hostel.

Lloyd advocates irrigating with three- (in this case not four-) inch aluminium pipes, laid out by hand across a paddock. Which is exactly what we do, only with four-inch pipes. 'I want to look after my back, Patrice, there's a lot of difference in the weight between four- and three-inch pipes.'

'What about the time and effort of shifting them every day?' I ask of a bloke who has moved thousands of pipes, thousands of times, across his paddocks during the decades he's owned Meadowdale.

'You've still got to set up an irrigator and drag it about. That takes time, too,' he says.

This is one of the problems with irrigation being used along the Pages and Isis rivers. People are investing in immense new infrastructure. A trip up Waverley Road these days has neighbours saying, 'Have you seen the size of those irrigators!' They're ludicrously large and, when turned on, can use up an annual water allocation in a matter of days.

Perhaps the legislation can make a difference. It's been proposed that once the river flow drops, we'll only be allowed to take a 'daily entitlement'. The sad joke is that some of those preposterous pumps will take that entitlement in a matter of minutes. Pump sizes and irrigation systems should match the river's capacity – unfortunately, around here they match the size of egos. Lloyd thinks it is all madness. His approach is a reminder that if you have a milkshake in a glass, you don't need a piece of garden hose as a straw.

With most cattle sold and no rain in sight, it's a good time to book a dozer and clean as many empty dams as we can afford to. First, Col brings over a sludge pump to slurp up the last of the muck. Then in goes the dozer to scrape out the sediment – and it is amazing how a mountain of mud comes out of even a small dam. At least they will be clean and ready if and when it ever rains.

This drought, the 2002 to 2003 drought, has finally hit the

headlines. Pictures of dying cattle are making front-page news in the broadsheets, each night the television carries another story about sheep being put onto failing wheat crops, emus descending from the ranges to eat crops, kangaroos being culled, the price of hay becoming astronomical, and the possibility of agistment now impossible with all of eastern Australia gripped by dryness. By winter we've a few young heifers strip-grazing on the lucerne in front of the house. As soon as they eat a strip, we fence off the area and irrigate. On the one hand, I want the lucerne to last; on the other I want it eaten off quickly, so we can irrigate again before the river dries up.

We'd been planning to sow more lucerne ever since we planted olives in the old lucerne patch, but never got around to it. Now, as the soil is cracking open, I make the bold decision to sow a new 10 hectares of lucerne before the drought moves into summer. We'll have to deep-rip the phalaris paddock and prepare it to sow. It's risky; if spring comes too early or is too hot, the lucerne seedlings could well perish. But I gamble we'll be able to irrigate it at least once and that should be enough to establish it, and, as soon as it rains, we'll get a proper crop. Even a little rain helps lucerne, whereas irrigating pasture just doesn't provide the same food quality or quantity.

I go down to the paddock and mark the area for a new electric fence and, within a couple of hours, Gavin is dragging the Agrow plough across the pasture, lifting the clumps of phalaris and exposing them to the sun to die. Within three weeks we're ready to sow the lucerne. By this stage the river has dropped considerably and each morning before checking email, sipping tea, or waking Aurora, I'm on the Internet site to see at what level the river is flowing. It's been fluctuating between 0.8 megalitres and 1 megalitre. The Draft Water Sharing Plan had set the 'cease to pump' level at 0.5 megalitres, so there is still some water left to take. Not that anyone is policing it. Everyone on the Pages and Isis is in full irrigation mode now. We're all aware that this drought is crushing and want our share of the water while it lasts.

Fortunately, some rain comes, almost half an inch, enough to settle the dust and bring on a germination in the paddock. We keep irrigating, dropping back the amount of time on each shift. Gavin moves the pipes twice a day and reaches the end when the river shows little sign of flow. The last flood, over a year ago now, left a sandbar that has reduced our waterhole for the pump by half. Although the gauging station tells me there's 0.8 megalitres flowing past Gundy, it's certainly underground at Elmswood.

There's nothing to do but wait and hope. The rush to sow means the paddock isn't the smooth seedbed we'd have liked. But then I can think of plenty of times when we've prepared the soil to perfection, only to have some other elemental disaster prevent a successful germination.

Waiting for rain can be all-consuming during drought, and isn't as dull as you'd think. When you first begin to suspect the dryness in the air could become a 'drought', everyone you meet has a theory and explains a different approach. This bloke was gambling on a short drought. That bloke was bold and sold everything. Another is handfeeding. They discuss which type of feed is best, how cattle from Tasmania are slipping, how old Herefords still look fat, how prices are low, and how many cows are being slaughtered.

After ten weeks, things are so bad, the road so full of cattle trucks, that people avoid talking about the drought. Living through it is enough. I see an upstream farmer in the bank and ask, 'Are you alright?' He looks like he has a terminal illness. 'Just drought-affected,' he says with a grim grin. And that is the truth of it.

Besides the grasses being eaten down, trees are shedding leaves, she-oak needles carpet the gullies, ancient post and rail fences up vertical hillsides are suddenly features in the landscape, and the geology of the hills (escarpments, fault lines, the folds and fractures of a million years) greet us everywhere we go.

16

It's a dark night in October 2002, and I'm sharing a Murrurundi paddock with hundreds of others, hugging myself and hopping from one foot to another trying to keep warm. We're all here in the cold, blustering wind for the inaugural performance, indeed the world premiere, of a play by local author Frank Wilson. This dogged, determined septuagenarian persuaded many people to help him create what he insists is 'Australia's first drive-in live theatre' – and he's organised nothing short of an epic. Called *Outlaw Mother*, the play is about Eliza Hall, mum of Ben, Australia's second most famous bushranger. Phillip will be the 'voice of history', Caroline from Gundy will play Eliza, and Royce, Aurora's bus driver, will be a Wonnarua tribesman.

The weather has been appalling all day and Frank, who spent months organising the event, was despairing of attracting an audience. Yet it seems that everyone in the district has come – with 4WDs backed up for kilometres. A bikie gang, all menace, leathers and Harleys, agree to do crowd control and are guiding arrivals to their parking spaces. If you choose, you can watch the production from inside your vehicle and tune into an FM radio station to hear the actors. But many people have braved the weather and are 'camping out', sitting on rugs or folding chairs in front of a bulldozed stage covered in lights and flapping canvas. The cast (and it's enormous) is rushing around full of nervous excitement, buttoning up each other's costumes and trying to remember their lines. It's a scene of partially controlled chaos, given that it's been nearly impossible to arrange a rehearsal due to the company being made up of farmers, miners, housewives, shop assistants, school children, teachers and old-age pensioners. So there were only a couple of non-dress rehearsals in a local hall, and most of the cast have never acted in anything. Ever. Nevertheless, judging by the size of the audience and their applause, the show is a success.

The heroine of Frank's play, Eliza Hall, was born in the slums of Dublin in 1807, spent some time in jail and was eventually transported to Australia for seven years, arriving in 1830. Four years later she married Benjamin Hall and in 1837 gave birth to son Ben – Ben the bushranger.

Thomas Haydon, proud owner of a property called Bloomfield, decided to subdivide a paddock beside the Pages into housing blocks. The result would be a small village, modestly called Haydonton. The first block, of two-and-a-half acres, went under the hammer in 1842 and was purchased by Ben's mum and dad. Here they built a three-room slab hut with glazed windows and a bark roof. They kept a tiny dairy herd, grew a few vegetables and picked fruit from their tiny orchard, selling surplus supplies to the district.

Ben left home as a teenager, robbed coaches, was charged with theft, and engaged in a sort of guerrilla warfare against officialdom. Regarded as an antipodean Robin Hood, he was both famous and infamous, and eventually was shot in 1865.

Eliza was illiterate when she arrived in Australia and probably remained so, yet she had regard for official documents and insisted that the deeds of the Halls' next home, at Blandford, be put in her name, otherwise she wouldn't move. Years later, when her wildly unreliable husband, who kept disappearing for years on end, decided he wanted to sell it, she put an advertisement in the local paper announcing that she and she alone was the rightful owner and that it bloody well wasn't for sale. So for seventeen years Eliza lived as a pioneer and matriarch at the head of the Pages. She's buried in St Joseph's churchyard at Murrurundi, under the name of Elizabeth, rather than the Eliza she was christened. Her death certificate shows she had eleven children.

Despite her criminal past and the recalcitrance of her husband, Eliza's feistiness and pioneering feminism earned her the respect of many of the local women, up to and including the wives of the landed gentry. They might have been far better off financially, but were nowhere near as free and spirited. Here, some 140 years later, was Eliza's community celebrating her life.

Sitting in a chair that had once belonged to the bushranger himself, Phillip read the narration. As his voice boomed out into the darkness, Caroline from Gundy went through Eliza's paces and, surrounding her on stage, were many of the women from Murrurundi – some of them grand-daughters and great-grand-daughters of the women they were portraying. In Murrurundi, the past and the present are close neighbours.

Whilst the Halls were making history, more people took up land in Haydonton; and, on the northern side of the Pages, Murrurundi developed as a twin city. Decades later a bridge joined the two towns and in 1913 they were amalgamated. The Haydon family still owns Bloomfield, part of the original land purchase, and they have the distinction of being the ANZ Bank's oldest clients.

Fortunately, the Haydons were all good correspondents and journal keepers and their archives are comprehensive and vivid. Peter Haydon, the present occupant, agrees to give me access and I spend hours in the little room, overlooking the Pages, where the family has stored nearly 200 years of their history. The details are mesmerising. Particular meals are recorded and the constant toing and froing of visitors – reminding us that the 19th century was a time of great social intercourse and mutual assistance. I read of different family members riding to the neighbours to 'take tea with the Bickham girls' or 'look at Whites' washpool' or just 'go to the Page'.

Included in the archives are letters written back to Ireland at a time when paper was in short supply, so the elegant hand-writing criss-crosses every sheet, vertically and horizontally and on both sides. Thus, in 1839 Thomas Haydon writes home, proud of the natural attributes of his new land purchase,

> I'm happy to inform you that I am now becoming a new settler having got from my brother one of the finest farms on the Pages River. On almost every farm they have to sink wells to obtain water, but I have the most abundant supply in the driest season having in the garden a fall of water about four feet high which I don't think can be equalled in the colony for sheep washing.

Thomas' water supply is still supplying the Bloomfield property. When I visit the pool at the height of the 2002 and 2003 drought, it's sustaining not only human life, but providing a refuge for platypus. However, Peter, who's been swimming there his entire life, says that like most of the river, it's slowly silting up.

While the adjacent land has been cleared, a handful of ancient white cedars grace the paddock and a few she-oaks cluster at the river's edge. The only ghostly traces of a tenant-farmer scheme are old robinias and English elms near the yards.

Diary entries in 1838 express disappointment: 'Birds without song, flowers without smell, trees without fruit.' Frustrations: 'Shot an opossum that was eating the grapes.' And gratitude: 'Although there's not much water, there are plenty of ducks, quail, pigeons and turkeys etc.'

Blandford, once known by the evocative Irish name of Donnybrook, was surrounded by vineyards owned by German settlers, of which no evidence survives, but older locals recall a time when the Emirates Park horse stud was covered in vineyards. Agriculture was more diverse back then, with corn, oats, grapes, fruit, wheat, cattle, sheep, horses and pigs. 'He shut up some time ago twenty-five pigs for breeding,' records Thomas Hayden in an 1839 letter. 'But they have all gone wild.'

Whilst wool and beef production provided solid returns in the early years, horses became increasingly important to many landowners. Peter reckons he has the oldest continuous Australian stock horse stud in the country, with bloodlines going back to the stallions and mares purchased by his antecedents. As early as 1834, horses from the Upper Hunter Valley were being despatched to outposts of the British Empire: to Madras to help crush the seditious Sikhs; to South Africa to attack the Boers; and in the Middle East, Australian stock horses (then called Walers), especially those bred at the head of the Pages River, acquired legendary status in the Battle of Beersheba on 31 October 1917.

Desperate to retake Gaza, the British hatched a strategically complex plot to seize Beersheba, thus securing water supplies. Lieutenant General Sir Harry Chauvel led his 15 000 mounted units for 30 miles one night, broke through Turkish forces and captured the town. Considered one of the epic feats of our military history, the battle has been depicted in two feature films, *40 000 Horsemen* (1940) and *The Light Horsemen* (1988).

Peter's ancestor, Lieutenant Guy Haydon, who'd survived Gallipoli, took part in the charge. Midnight, his dark brown mare, was killed jumping a trench, but Guy lived to write extraordinary letters about the front line, while convalescing in a Cairo hospital.

> We trotted for the first two miles, then the Turks opened fire on us . . . we could hardly hear anything for the noise of their rifles and machine guns. As soon as their fire started we galloped and you never heard such awful yells as our boys let out. They never hesitated or faltered for a moment. It was grand. Every now and then a rider would fall off or a horse fall shot, but the line swept on . . .

Stock horses are a dense, muscular breed of about 15 to 16 hands. Despite exporting over 100 000 for World War I and thousands more for hard work and recreation, it wasn't until 1971, when a group of 'enthusiasts' met at the Scone Royal Hotel, that a society was formed and the breed was officially recognised as Australian. Like many aspects of Australian life, including dogs and human beings, the Australian stock horse has a 'bitzer' heritage.

International interest spread after their prominent role in the Olympics Opening Ceremony, sending the price of stock horses soaring. Today the Scone headquarters registers 150 000 horses – and at least 600 mares live in the surrounding hills. Unlike the thoroughbreds of the racing industry, the Australian stock horses are a democratic, down-to-earth breed.

Turning the pages of the rainfall and temperature charts collated over the last 100 years, I see that droughts usually break just before, or just after, Christmas, with big follow-up rains falling in

February and snow the next winter. I'd also been reading local history and realised that three-year droughts were all too common. The December 2002 rain half-fills some of the cleaned-out dams and triggers the growth of native grasses. Upstream, at the head of the Pages and the Isis, they had a lot of rain – making the floods faster than expected. This would be the first time there'd been a surge of water over the snag created by the death of the she-oak and the manoeuvring skills of Terry and Col. In place for almost a year, the she-oak had merely lain beside the bank, the dying needles falling off as they were replaced with a crop of turtles. Sometimes we'd see a whole row of them sunning themselves, like plates on a dresser.

Many of the native trees we'd planted on the bank had died and the survivors were only just clinging to life. Surrounded by dried weeds, the area looked like a wasteland rather than the successful stream-bank rehabilitation effort I'd planned. So a flood, a minor flood, was just what we needed. And when it came, I wasn't measuring it against the bridge as usual – but waiting to see the effect of the snag on its flow and hoping the seedlings would be briefly submerged. If the flood was too strong, the tree could slam dangerously against the bridge or simply be washed downstream.

Removing snags from rivers has been central to management theory for years. To improve navigability and minimise flooding, snag boats on the Murray removed three million trees. By the 1970s only a thousand or so were left – at which time scientists realised snags were invaluable in minimising river-bank erosion.

Snags are now considered equivalent to marine reefs, because they provide a habitat for fish and plants, bacteria, fungi and microscopic invertebrates. More snags equals more biodiversity. Even the slimy coating on the old wood, which feels disgusting when you stand on it, has a role to play. It turns out to be a 'bio-film' providing a rich food source for invertebrates – as well as trapping nitrogen and phosphorus, which, in turn, helps prevent algal bloom. Excited by a theory that actually seems to work, we spent hours down by the river watching new life gather around the newly created bank.

Unfortunately, carp survived the drought. Trapped in isolated pools, sometimes just a few centimetres deep, we'd expected any day to see them belly-up. Phillip tried to shoot a few, but even a sheet of water was enough to deflect a 22 bullet. In fact, the carp seemed to accommodate themselves to the shallowest waters, which had the effect of making them look even larger. No wonder they flourish. Here's a fish you can't catch, shoot or starve. The good news was that catfish and turtles were back sharing the water with them.

To the delight of Aurora's dog, Tommy, there was an immense proliferation of insects. Uninterested in rounding up sheep, he spends all summer in pursuit of dragonflies. Crouching near them, pointing his nose at them, whiskers quivering, he slowly inches towards them and leaps! Absolutely uselessly. Though we've never seen him catch a dragonfly, he's never discouraged. The dragonflies seem to enjoy it too. You'd swear they were teasing him, tantalising him with their proximity, sometimes orbiting his head.

For Tommy, like Aurora, the river is heaven. While he tries to catch one of these little helicopters, Aurora scoops up water laden with dragonfly larvae. Other larvae attach themselves to snags, so they can sift the passing water for nourishment. I'm delighted that our snag is being colonised and hope the river's flow is sufficient to sustain these minuscule communities.

Downstream, 300 metres from the bridge, is another dead she-oak, dumped by floodwaters years ago. Here immense amounts of gravel were removed by the council to help it bituminise Gundy Road back in the 1980s. Although the crater was bulldozed right into the hyporheic zone, it has long since filled and the river has formed a deep channel on the eastern bank. I've always wanted to move the tree, because it looks so damned ugly, but now it would be illegal to do so. That's the prerogative of future floods. Yet I'm hopeful one day the tree will be swung against the river bank where, like our sensitive, new-age snag, it can become a haven for life.

The DLWC suggests that irrigation licences on the Pages and Isis be traded downstream. This way, less water upstream will be extracted and, hopefully, more water will flow into the Segenhoe aquifer.

Sadly, most of the licences traded over the past two years have gone in the wrong direction. The shifts in water are notified in the local paper and written objections are invited. When I phone to query the situation, I'm told that, despite the philosophy of wanting to move water extraction downstream, it is true that, yes, licences are moving up. If that's what money wants, that's what money gets. Money not only talks; clearly, it has the last word.

'Yes, madam, you can complain about where water trading is heading, but my organisation is only interested in how an application might affect your business – directly.' Otherwise shut up.

Another issue that's raised at many a meeting and written about in much of the official literature is that some soil types are absolutely and utterly unsuited to irrigation. To pour precious water onto such acreage is an outrageous waste of a resource. Even if it means that, in the future, marginal land with dead soils will be watered extensively and expensively, leaving good, agricultural acreage downstream dying of thirst, so be it. If you can afford to buy the licence, the equipment, and pay the labour, you can irrigate anything anywhere any time.

With irrigation destined to increase, how will Segenhoe Valley ever be able to recharge its massive aquifer? Since it depends on the ravaged river system, it's incomprehensible that the aquifer could be 'five times overallocated' with licences in the first place. So I ask Garry from the DLWC if he's sure about the numbers.

'We know the rainfall figures for the past 100 years and we theoretically know the infiltration rates of the soils around Segenhoe. By modelling we can work out that the long-term recharge of water into the aquifer is about 1000 megalitres per year.'

Yet the DLWC has issued licences that add up to 5000 megalitres.

There are many questions that don't seem to have answers. Are all the soils the same? Where are the underground paths to and from the river? Does it connect to nearby aquifers? Is it connected to Lake Glenbawn? Because of these 'unknowns', as they're categorised, there won't be a reduction or 'claw back' of licences. On top of this overallocation, anyone in Segenhoe can buy surface-water licences from upstream and take more water, which is meant to be recharging the aquifer, from the river. The 'unknowns' are what everyone who despises any attempt at water reform uses to justify their continuing irresponsibility.

Ironically, despite all the water Segenhoe draws to keep its grounds looking as pretty as a picture, the place may not be all that healthy for those million-dollar thoroughbreds. 'This country looks good for horses, but it's not, if you know what I mean,' says one vet. Like pavlova, I suggest – luscious, but it usually makes you feel sick. 'Exactly.'

It's been said by many around here that the 'classic breeders' have long left the thoroughbred industry – that it's no longer an art, but a science. And perhaps a pseudo-science. Breeding is planned on paper and not enough attention is being paid to what is actually happening in the paddocks. What I'm told (in whispers by vets who dare not publicly complain) is that things are going terribly wrong.

With huge money being paid for yearlings and two-year-olds, horses are being forced to grow too fast. According to vets, around

30 per cent of horses are born with leg problems, or, to use the official term, orthopaedic bone-development disease. There's a lack of bone mineralisation, so the horses get weak bones and collapsing cartilages. Some will self-correct, others need surgical attention. In other words, all this thorough-breeding is leading to weaker animals whose deformities may need to be surgically corrected before they can be sold. So bad is the problem that some yearling sales now X-ray horses to try to ensure the horse will be able to stand up properly, let alone run in a race. But just as some people can beat a lie detector, others can beat these X-rays. Corrective surgery techniques are so sophisticated that, like Californian plastic surgery, they can conceal all sorts of problems. A good surgeon can operate without scarring the animal, so that no sign of the past defect is detectable, not even by X-ray and, least of all, by a prospective buyer. This reminds me of similar problems in the intensive poultry business where chooks, heading for the rotary roasters, are forced to grow so fast they can't stand up.

Most foals are born between 55 to 65 kilos, and within four months are expected to weigh 300 kilos. A horse doesn't naturally mature for four or five years. 'They have to have the right nutrition to make that sort of weight gain and irrigated pastures aren't always enough,' says one vet. Pasture samples at one Segenhoe horse stud found that, despite regular fertilising and hundreds of megalitres of water, the protein level of the feed was okay, but lacked phosphorus, copper and numerous trace elements. They had to provide all horses with mineral supplements. Irrigation can't deliver the required diet, despite the paddocks looking so appetising.

Of course, you can't beat green grass. During the drought when we were selling stock – or giving it away – buyers always asked, 'How much "green feed" is left?' Green grass provides more than nutrition, it provides meat with flavour. But in the Segenhoe Valley, the grass on the other side of the fence can be too green – cosmetic green. One vet is lethally cynical: 'These places are irrigated for shareholders, not horses.'

Farmers without a skerrick of green feed drive through the valley past the studs and are bitterly resentful at the water being sprayed around. We know it's not coming from the Pages River because it's dry, so could all this water be coming from the aquifer?

Two of the biggest studs have bought water security from the Hunter River and buried kilometres of pipe leading to their studs. Another has secured a licence to pump direct from Lake Glenbawn. One stud bought a staggering 900 megalitres, such was their concern that the Pages aquifer and the pending water reforms mightn't guarantee their irrigation.

It's a local joke that the forests of the coast have been logged simply to fence in the horses at Scone. And it's a fact that the studs acquire a disproportionate amount of the state's finest timber. Fencing is another way a horse stud advertises itself. Where other farmers struggle to pay for a few uprights and some barbed wire, the studs' fencing is invariably ostentatious – the most posh use strainer posts (medium-sized trees cut to length) for every post, not just split timbers. Wide-sawn timber horizontals don't rest atop the strainers, but are cut into them. Most studs then paint the railings with sump-oil to blacken them and a hot-electric wire goes around the whole lot. In many cases the ground beneath the fence is poisoned with herbicide, so the final effect is of absolute neatness.

The sport of kings and queens, sheiks, tycoons and shock jocks dominates our area. Anyone thrilled to see horses in their wild state must, surely, have mixed feelings seeing them fenced in these glamour gulags for the gambling industry. And sometimes it seems that the horses don't like it either. Recently a multi-million-dollar stallion killed itself by galloping into a Segenhoe fence. Many weren't surprised.

Horse studs are really part of the suburban sprawl, bringing with them a plethora of offices, stables, arenas and staff residences. Most studs begin by felling every unphotogenic tree and poisoning the native grasses, thus killing any hope for biodiversity. After applying herbicides, they sow the new, high-protein grasses and legumes, in contrast to farmers I know who lock up paddocks for decades, or rotate stock, in the hope of re-establishing native grasses.

More and more of the district is being developed by money from all over the world – from Asian millionaires and oil revenues from the Emirates. How much land previously employed for food production should be used by the thoroughbred industry? How much of our precious water should be channelled into it? Money silences the questions and provides the answers. I protest that *the land itself should dictate its use*, but this idea appears to be hopelessly romantic.

Horse breeding is economically advantageous, but environmentally unstable. The Pages is unable to meet the increasing demands of horses with bad legs. There'll come a time when nature will show us that it can't take it any more. The pastures will take no more. The poor horses will take no more. But we're not there yet, not quite. The studs proliferate and every new one has effects beyond its fences – making it hard for neighbouring land to be used for anything else.

Scone is proud of its horse industry and the horse has become its corporate symbol. There's a well-loved statue of a mare and foal in a local park, providing countless photo opportunities; there are annual horse parades and an increasingly successful local racetrack. We're told again and again that this is an 'elite' industry. But behind that word 'elite' lurks a policy of poisoning the earth, regarding land as a mere commodity and – not deliberately, but incidentally – breeding deformed horses.

The Pages and Tributaries Water-Users Association seems to have evaporated during the drought. Our last meeting for 2002 is adjourned. Then, as the mood at Elmswood and all around descends into despair, rain trickles down. 'It looks like we've called in the carpet layers,' says Phillip, as the paddocks turn green overnight. Our beloved Kewell Creek flows again, and Phillip and Aurora splash beneath the little waterfall in their secret rock pool. The Pages expands encouragingly, although it never covers the bridge.

The DLWC water restriction didn't matter after all. The water begins to clear and once again there's life around and in the Pages River. But no sooner has Aurora taken the decorations off the reusable plastic Christmas tree than the drought returns with a roar – and the river shrinks back to stagnant pools. Soon Gavin is turning off our mono-pump every few hours to give the diminishing waterhole some time to refill with seepage. Downstream, at Gundy, household pumps are gasping on dry gravel. The pumps are breathing air, despite the 300 millimetres of rain that fell at the head of the Isis catchment and the 200 millimetres that drenched the head of the Pages. Where has all the water gone? Here we are, after four years of discussion and negotiation, about to enter a time of brawling, accusations and abuse. So much for self-regulation.

At the request of downstream water-users, I email the DLWC suggesting a cease-to-pump ruling is, once again, due for consideration. For days this request is caught up in a ping-pong match between different departments. Officials ask if we've agreed to a cease-to-pump yet. It's not a question of agreeing. We'll be having dry showers soon. Yet irrigation upstream from pools, mainly on the Isis, is in full swing.

What starts as a worried phone call from a Gundy resident leads to a week of growing rage with irrigators. I check the gauge reading on the Internet: the river is flowing at 0.3 megalitres at Gundy. That's already 0.2 under the draft unregulated water-sharing plans – and under what we'd considered reasonable. The Isis has also stopped flowing into the Pages. So I email the DLWC again suggesting they announce a cease-to-pump. Days later the DLWC says the Pages should be on cease-to-pump, but it will be sufficient for the Isis to be on 'restrictions'.

When I speak to the DLWC about this odd decision, I'm told the people up the Isis were peeved that we down at the lower Pages hadn't been on restrictions before Christmas, so they now feel entitled to keep pumping. I can't believe this nonsense. After years of trying to accept that the volume of water extraction was at the heart of the issue, here was an official giving the nod to one part of the river system – the one with the biggest pumps. They were to have preference over a couple of tiny farms irrigating near Gundy, where farmers had to disconnect, lug and reconnect some twenty pipes. Why should their entitlements be sacrificed to thirsty, greedy people upstream? It was even worse to learn that the DLWC had empathised with those pre-Christmas grizzles. It looked like the DLWC wasn't making decisions based on facts, but responding to pressure.

Scores of meetings, a tonne of reports, proudly announced policies, the promise of reform and of a new era for our poor old river had come to this. It had come to nothing. I went inside and typed up my letter of resignation.

> Dear Sir,
>
> After four years as secretary of the Water-Users Association, I resign. During this past week, as residents of Gundy struggle to get water, the department has

proved incapable of implementing a fair and equitable use of irrigation water. Having accepted the conversion of licences to volumetric, having tried to understand the draft principles, having accepted that 0.5 megalitres is the cease-to-pump level, having acknowledged that the residents of Gundy were to have a priority, having conceded that pools aren't to be pumped out (that to argue 'my waterhole is still okay' doesn't hold water), that when the Isis stops flowing into the Pages, that's more than enough, having heard everyone make promises they now refuse to keep, the DLWC sees fit to allow one section, one faction of the association to irrigate, and not another.

What conclusion are we to draw from this? Are we seeing incompetence or corruption? Cowardice or stupidity? Whatever the reason for this dereliction of duty, this abdication from a principle, this backtracking from negotiated agreements, the department should be ashamed of itself.

But I don't send it. Why bother? Clearly, nobody gives a stuff.

While most of us on the Pages River have been busy dealing with the drought, water restrictions and the manoeuvrings of Bickham Coal, another issue reaches a conclusion. One Thursday, at the end of 2002, I sit at the back of a room in a Muswellbrook motel listening to the Hunter River Management Committee sign off on the Wybong Creek Water-Sharing Plan. This plan has been the template for us on the Pages. Listening to Roland lobby for the fish, Bruce and Lloyd for the farmers, Marg for the environment and Macquarie Generation's Errol for electricity, along with a dozen others just worrying aloud, I realised every river's crisis is unique. However, whatever the variations on the theme, all arguments face the same elemental and profound reality: there simply isn't enough water.

There was little joy in the room and no sense of achievement as the final touches were being made to the plan before being sent to the minister. Errol, for one, is mightily unimpressed with the thought of floodwaters from the Wybong, or any other unregulated stream for that matter, being unavailable for his power stations. Finally, as Wej Paradice is about to close the meeting, Garry speaks, not wanting the meeting and the whole project to end with a whimper, 'This is an important day. After years of debate we've finally reached agreement on a water-sharing plan.'

Still there's no applause, no handshakes, no back slapping and, sitting beside me, Don, the President of the Wybong Water-Users Group, shakes his head. From the outset he's been upset that not one water user from the river was on this committee, saying, 'It's just not democratic.'

I'll never forget the summer of 2002 to 2003 with its searing heat. It was when our precious Water-Sharing Plan was undermined, and when the biggest potential user upstream, Bickham Coal, fell strategically silent.

Directed by the minister to 'reassess' the bulk-extract proposal, Bickham Coal undertakes to present a new plan in January 2003 and John Roberts manages to secure a meeting with Bickham to examine the geological proposition that we believe is fatally flawed. But after a five-hour drive to Singleton, he is denied the promised access to the data and drives home in a fury. If what we are saying is wrong, why doesn't Bickham Coal simply show us the information? After six months we've seen nothing. We prod the Department of Mineral Resources, the company and the consultative committee, and get nowhere. As the New South Wales state election of 22 March becomes the government's focus, all the major parties want the political embarrassment of Bickham Coal off the agenda and it becomes impossible to talk to anyone in Macquarie Street.

Nor does the issue instantly resurrect after the election. Carr's Labor Government gains a third term, the Greens double their vote just about everywhere and, in the twinkling of an eye, the DLWC ceases to exist. So, in the middle of the Great Water Debate, we get a conjuring act – a disappearing department! Like all staged illusions, it's done with mirrors. The two ministers previously responsible for water are relegated to the backbench. Staff are reshuffled, desks rearranged and a fortune spent reprinting letterheads. Then off we go again. To tear a department apart after exhausting everyone by means of the so-called 'process' is crazy. More importantly, it leaves the community demoralised, cynical and bitter. Consultation starts to look like a hoax.

After clawing back some licences across the state, haphazardly restricting water flows, separating land ownership from water licences, it's like a game of Snakes and Ladders, without the ladders. We've been driven down a dead-end road into a swamp. Now we have to persuade a new minister and a new Sir Humphrey. We must find the energy to start all over again.

Weeks pass and we learn that a new bureaucracy will be created called the Department of Infrastructure, Planning and Natural Resources (DIPNR). The Dip – let's hope it dips in the right direction.

Phillip is sitting in the Main Library with Aurora, sifting through a thousand knock-knock jokes for our latest joke book.

'Knock, knock,' he says.

'Who's there?' I wearily ask.

'It doesn't make any difference. You're stuffed anyway.'

17

Having now walked along much of the course of the Pages and witnessed its moods, it's clear to see how a few simple changes – such as removing introduced stock just some of the time – could re-create a magnificent diversity. I've also seen how well-intentioned interference can make things worse. Although I'm not sure if it's true, I want to believe that people are learning to love the river more, rather than demanding its obedience.

Sometimes I've paced the river's edge counting steps, bypassing the ordinary, in search of the dramatic. I'd want the flow of the Pages to be more grand, offering movement on a still day, but not be so fast as to be dangerous. To see more native fish, native animals, native trees, native wildflowers, native raspberries to pick, but mostly longing for our river to be more significant.

Mighty rivers, famous rivers, rivers that flow through history carry as many stories as silt, flooding the imagination.

Here I am, in contrast, walking beside a river that few have heard of and even fewer have seen – and yet the *more* I've got to know it, the *more* it seems to matter and the *more* determined I am to care for it. The Yangtze has a greater flow in a moment than the Pages has in years. The Danube matters to the people of many countries. Nations will go to war over the Mekong, as they will over the Euphrates, but I'm swept up in the politics and plight of a river whose likely disappearance will probably pass without notice.

You can't even row a boat down the Pages.

'The end of the Pages,' that's what someone says to me this morning. 'The coal mine will be the end of the Pages.'

The Pages meets the Hunter River just ahead of me – and that *is* the end of it, the true end of it. It's not a beautiful end amidst quiet beauty. Instead there's the head-banging sound of a gravel quarry that's been operating beside the Hunter in an old riverbed for years. It's a moment of the pastoral meeting the industrial.

Turning the last page of the Pages, I find more flood-mitigation work, profoundly ugly erosion, another rubbish tip, some areas devoid of cattle and others heavily overgrazed. A solitary Hereford gives me a ponderous gaze and hundreds of corellas lift from the branches of the she-oaks, screeching and circling around to land again after I pass.

I reach the Allan Bridge, the last of the classic wood-and-cable bridges, the others having fallen victim to floods, overladen trucks or municipal neglect. Not so long ago these bridges crossed and recrossed the Pages and the Isis, white-painted and charming, covered in loose boards that made a xylophone sound beneath a vehicle's wheels. They've been replaced with serviceable, no-nonsense concrete bridges that can survive the pounding of floods beneath them and the hammer of timber jinkers above them. Pity that they're utterly graceless.

Here our river ends without dignity or poetry. The Hunter

swallows the Pages and moves on. Across the river from the quarry, expensive fences are being erected as the thoroughbred industry extends its influence. From here the flow is under human control, channelled and directed by the engineers of Lake Glenbawn. You can't cross it on foot today; the sanctioned flow is too fast. Downstream, it regains some beauty. Fringed by willows, watched by cows nestled in couch grass, the Hunter could be the Thames flowing towards the Henley Regatta.

F. Scott Fitzgerald ends *The Great Gatsby* with one of my favourite lines, 'So we beat on, boats against the current, borne back ceaselessly into the past.' Now our river is beaten back by the currents of a human future. The destiny of the Pages River – my river – lies in our hands.

ACKNOWLEDGEMENTS

I'm indebted to everyone who lives along the Pages River – and many along the Isis – who took the time to show me their part of the river and offer leads to the stories told. In particular, to everyone in the Pages and Tributaries Water-Users Group who, unbeknown to them, triggered the idea for this book.

In particular, thanks are due to three women who are keeping alive the history of the Upper Hunter Valley: Audrey Entwistle, Una Price and Barbara Riddell.

Thanks also to: Peter Haydon, beneficiary of one of the great family history collections; Scone Branch of the New South Wales National Parks and Wildlife Service; Garry Hunt and Paul Collins, formerly of the DLWC, now both working for the New South Wales Department of Infrastructure, Planning and Natural Resources; Danny Lewer, who started and finished the Lower Pages Rivercare Plan; Wej Paradice, for allowing me to see the workings of the Hunter River Management Committee; farmer Lloyd Rossington, who knows and loves water; Mary Steepe; Gerard McLaughlin, from the Scone office of the Rural Lands Protection Board; Al Bashford, librarian at the New South Wales Department of Mineral Resources; Al Simpson; Travis Peake; Manny Bloomfield; John Simpson; Cameron Archer;

Michaela Malone; Helen Brayshaw; Ann McMullin; Rick Wright; John Roberts; Jim Comerford; Brother Henry; Dr Andrew Bolton; Peter J. Hancock; Gavin Prescott; Colin Watts; Yvonne Mitchell; Elizabeth Bullen; Gail Cork; Craig Dickinson; Deirdre Provis, who read parts of the manuscript and advised; and at Penguin Books, Heather Cam, Meredith Rose, Nikki Townsend, Leanne Marcuzzi and publisher Julie Gibbs, for thinking the idea was worthwhile.

Most of all to Phillip and Aurora, who shared it all.

FURTHER READING

Ball, Philip, *H₂O: A Biography of Water* (Phoenix, 1999).

Bonyhady, Tim, *The Colonial Earth* (The Miegunyah Press, 2000).

Brown, Lester R., *Eco-Economy* (Earthscan Publications, 2001).

Daly, Herman E., *Beyond Growth: The economics of sustainable development* (Beacon Press, 1996).

Fullerton, Ticky, *Watershed* (ABC Books, 2001).

Gray, Nancy, *Wilfred Green of Gundy and the stories he told*, Scone Historical Monograph No. 6 (Scone and Upper Hunter Historical Society, 1979).

Lewer, Daniel, *Lower Pages River – Hunter River Confluence – Isis River Confluence Rivercare Plan Companion Booklet* (Scone Landcare Group Inc., 2002).

Peake, Travis, *Hunter Bushland Resource Kit* (Hunter Catchment Management Trust, 2003).

Robèrt, Dr Karl-Henrik, *The Hidden Leadership Towards Sustainability* (Institute of Environmental Studies, UNSW, 2001).

Scone and Upper Hunter Historical Society, *Journal* (all issues) (Scone and Upper Hunter Historical Society, various dates).

White, Mary E., *The Greening of Gondwana* (Reed Books, 1986).
— *After the Greening* (Kangaroo Press, 1994).
— *Listen . . . Our Land is Crying* (Kangaroo Press, 1997).
— *Running Down: Water in a changing land* (Kangaroo Press, 2000).
— *Earth Alive: From microbes to a living planet* (Rosenberg Publishing, 2003).

Wood, W. Allan, *Dawn in the Valley: The story of settlement in the Hunter River Valley to 1833* (Wentworth Books, 1972).

Wright, Judith, *The Generations of Men* (Oxford University Press, 1959).

INDEX

A

Aborigines
  artefacts 75–76
  'bool', beverage 47
  carved trees 72
  Murrawin tribe 2
  Segenhoe 47
  significance of Burning
    Mountain 182–183
  Wonnarua 1
Adams, Phillip
  leaking dam 99
  libraries 133–134
  moving carcasses 176
  rescuing goldfish 170
Agaricus campestris 96
Agriculture, Dept of, NSW 186
Alectryon oleifolius 173
algae
  Aphanizomenon 138–141
  Cladophora 120
Alliance for Bio-Integrity 193
Angophora floribunda 61, 87, 172
Argentina 54–55
Australian Farmers' Attitudes to Rural Environmental Issues 1991–2001 188

B

Bathurst burr 39, 135
Beersheba, Battle of 215–216
Ben Hall Mountain 16
Bengalla Coal Mine 90–93
  versus Rosemount 66
Bersten, Ian 13
Bickham
  graves 60
  property 59
Bickham Coal Mine
  exploration 62–67
  gagging debate 93–94
  political agenda 230
  Review of Environmental
    Factors (REF) Oct. 2002,
    142–144
  water report 142–145
BHAG (Big Hairy Audacious
  Goal) 177–179
birdlife 113–114
blackberries 17–18
Black Mountain 118, 158
Blandford
  Ben Hall's home 212
  Donnybrook 214
  history 214
Bloomfield, Manny 128–129
biodynamics
  dairy 202
  water 23
Bobbies Reserve 69–70
Bonyhady, Tim 178
Boort 192–196
Brayshaw, Helen 71–76
Brown, Bob 93

Brushy Hill 68–70
Brushy Hill Dam 68–70
Burke, Governor 46
Burning Mountain 182–183
    Nature Reserve 74
butterflies 185–186

C

Cameron's Gorge 35
    Aboriginal artefacts 75–76
    dam 67–71
    walk 72–76
Capuchin Hermitage 13, 138
Capuchin Franciscan Order 13
Casuarina cunninghamii (she-oak)
    admiration for 100
    chainsawed 165
    fallen over 39–42
    leaves falling 209
    lopped for feed 52
Carboniferous Period 118–119
cattle 121–127
    drowning in dams 171, 176
    grazing on river banks 134–135
    see also drought
chemicals 135
Cladophora algae 120
coal
    description 31, 129–130
    discovery in the Hunter 130
    Greta seam 131
    to pay for the environment 190
Coal Acquisition Act 66
coal mining
    Bengalla Coal Mine 91–93
    fossil industry 93
    history 183–185
    near rivers 190
    overburden 190
    tax 90
    see also Bickham Coal Mine; mining
Comerford, Jim 132–133

committees 186–189
commons 162–166
    carriers of ideas 137
community consultation 142–148
confluence 82
Council of Australian Governments (COAG) 32, 177
Cranky Corner Land System 67
Cunningham, Allan 24
Cunningham, Peter 24
Cupricide 148
Cymbidium canaliculatum 173–174

D

dam, farm
    building 98–102
    cleaning out 206
    where to build 85–86
dam, municipal
    attitudes to 22
    in Hunter catchment 24
Dangar, Henry 19, 160
Donnybrook (early name for Blandford) 214
dredging 26–30
drought
    cattle 51–52
    defiant cow 121–127
    effect on the river 115, 120
    olives 53
    sowing lucerne 207–209
drowning in Pages 9–11
Dumaresq, Henry 24
Dunn, Phillip 139–140

E

echidna 18
Edgeworth David, T.W. 131–132
Environmental Protection Agency (EPA) 162–164
Environmental Protection and Biodiversity Conservation Act, 1999 177

erosion 25–27
Eucalyptus melliodora 5
European Union (EU) subsidies 199

F
farmers' attitudes to the environment 188–189
fencing 47, 48
fish
    bass 71
    carp 3, 166–167
    catfish 3, 166–167
    in general 167–171
    goldfish 170
    introduced species 168
    mullet 48
    native fish 167–169
    trout 171
fishing 167
floods
    December 2003 217
    Elmswood 6–9
    flood of 2000 149–155
    and genetic engineering 155
    Hunter Valley 22–30
    mitigation, Segenhoe 49–50
food, price of 198
Foot and Mouth Disease 126
Foster, David 62, 128

G
geology 116–120
Glenbawn Dam 23
goats 17–18
Greens 93–94
Gundy 7, 22
    Back to Gundy 71
    Gourmet Food Fair 175–177
    water shortages 226–228
Gwydir Grove 198

H
Hall, Ben 210–212
Hall, Eliza 210–212
Haydon, Lt Guy 215–216
Haydon, Peter 213–216
Haydon, Thomas 211, 214–215
Haydonton 211, 213
Herbertson, Dr J. 145–146
herons, white-faced 170
Hibberta scandens 16
High Valley 13–18
High Valley Falls 15, 19
Hodges, Peter 164, 180–181
Horse Parade 175–177
horses
    Australian stock horse 21, 215–216
    breeding 221–223
    fencing at studs 223–224
    Segenhoe 47–50
    water use 220
Hunt, John Horbury 61
Hunt, Garry 37
Hunter Catchment Management Trust 38, 173
Hunter River Management Committee 39, 229
Hunter River rehabilitation 89–92
Hunter Valley Conservation Trust 27
Hunter Valley Research Foundation 28, 38
hyporheic zone 115

I
Infrastructure, Planning and Natural Resources, Dept of, NSW 231
irrigation 204–209
    equipment 205–206
    no nutrition guarantee 222–223
    on marginal land 220
Isis River
    flood 217

irrigation pumps 206, 227
map 18
water restrictions 226–228

J
Joseph's Rest 16

K
Kamilaroi Plateau 16, 17
Kingdon Ponds 159
    geology 160
Kewell Creek 23
Keys Bridge 89

L
Land and Water Conservation,
    Dept of, NSW 32, 34, 41, 49,
    84, 88, 230
Landcare 188
legislation *see* water reform
lime mine 68
Liverpool Ranges 25
Livingston, Alexander 25
Lake Glenbawn Dam 23
latitude and longitude (confluence) 82
Local Agenda 21, 180–181
lucerne 201–206
    sowing at Elmswood 207–209

M
Maddox, Sir John 184
maidenhair fern 97
Maitland Mercury 20
masked lapwings 169–170
meetings, committee 186
McDougall, Ken 192–196
McIntyre, Peter 45, 47
Mineral Resources, Dept of, NSW 62
mining
    general 183–185
    lime mine 68

    see also coal mining
Monteith, N.H. 203
Moriarty, Mr E.W. 25
Mount Constable 16
Mount MacKillop 16
Mount St Francis 16
Mount Torreggiani 16
mullet 48
Murrawin tribe 2
Murrurūndi
    buildings 61
    drive-in theatre 210–213
    ford 19
    mining history 183–185
    twin city to Haydonton 213
    water supply 138–141, 148
mushrooms 96–98

N
Natural Resource Management
    Committee (Hunter) 188
Natural Step, The 145–146
nature reserves
    Burning Mountain 74, 182–183
    Cameron's Gorge 35, 67–71, 72–76
Newcastle 24
    Harbour 26–30
Notelaea microcarpa 156

O
oil-shale rock 184–185
olives
    in Argentina 54–55
    Australian Industry 191–200
    at Boort 192–196
    Elmswood Olive Grove 54–55
    Gwydir Grove 198
    harvest 109–113
    irrigation 108
    in Middle East 56
    Olivecorp 192–195

symbolism 57
Viva 198
Olivecorp 192–197
Outlaw Mother 210–213

P

Pages and Tributaries Water-Users Association 33, 35, 225, 227
Pages Creek Road 12
Pages River
    bank stabilisation 49
    beauty 23–24
    clearing 24–25
    dam 22
    descriptions 1, 2, 20, 21
    drought 115
    end of 233–234
    erosion 25, 88
    expectations of 101
    floodwater 22
    funding river works 88–90
    historic vegetation 24, 45
    Management Plan 36
    naming 18–19
    overgrazing 203–204
    sheep 25
    river works 88–90
    tip 77–78
    water volume of licences 83
    width 28
    white settlement, Segenhoe 45
Paradice, Wej 38, 229
Parkville Piggery 159–164
Peabody Resources Ltd 90–91
Peake, Travis 171–174
pigs 159–164
platypus 40
poplars 49
Potter McQueen, Thomas 19, 45–50
prickly pear 52
private irrigation schemes
    attempt at Segenhoe in 1890 102
    possibility of 86
    in Scone Shire 103–104
pumps
    at Elmswood 6–8
    on Waverly Road 206

R

raspberries, wild 48
rehabilitation
    of Hunter River 89–92
    of Pages River 89
    see also water reform
Rhiannon, Lee 94
Riddell, Barbara 19–20
river works, funding of 88–90
Rivercare 44–45
Roberts, John 117–119, 143, 160, 230
Robèrt, Karl Henrik 145
rogaining 79–83
Rosemount Wines 66
Rossington, Lloyd 204–206

S

Salix babylonica 29
Scone Council
    Development Committee 179
    Local Agenda 21 180–181
    piggery 163
Scone Horse Parade 175
Scotts Creek 23
Segenhoe Valley
    aquifer 220–223
    history 45–47
    horse industry 221–225
    in the 1950s 89
    irrigation 219–225
    position 68
    today 47–50
she-oak (*Casuarina cunninghamii*)
    admiration for 100

    chainsawed 165
    fallen over 39–42
    leaves falling 209
    lopped for feed 52
'Shiny Pants Brigade' 32
Shiralee, The 71
Shortland, Lt John 130
Singer, Peter 159
snags
    general 217–219
    she-oak 165
Southcorp 66
spiders 112
spinifex 68
Splitters Creek 23
Stephens, Robert (Yellow Bob) 69, 70
Stewart, John 72
studs *see* horses

T

Telstra 87–88, 189–190
Temi mine 184–185
Tertiary Period 119
Timbercorp 195
Travelling Stock Reserve 69
trees
    denuded 20
    history of clearing 203

U

US Food and Drug Administration (FDA) 193

V

Viva olive oil 198–199

W

Walers 215
Warlands Creek 23

wasps 94–96
water reform
    bureaucracy 33
    COAG 32
    financing 104
    legislation 177–179
    meeting at Gundy 42–44
    meetings, too many 186–189
    NSW Water Management Act 177
    prioritising issues 83–87
    R for rivers 31
    restrictions 226–228
    sleepers 83
    start of the debate 31–39
    trading 219–220
    volumetric conversions 42–44
WaterWise
    course 104–107
    decision 85
weeds 116
    along river 134–137
weeping willows 29–30
Wilson, Frank 210–213
Wilson, John 186–189
Wilton, Rev. C.P.N. 182
Wingen 182
Wingen Maid 183
Wonnarua
    occupation 1, 3
    sky god Biami story 182–183
Wonnarua Ridge 16
Wood, Allan 46
Wright, Judith McKinney, 59–61

Y

Yellow Bob (Robert Stephens) 69–71